Comprehensive Quality by Design in Pharmaceutical Development and Manufacture

Topical Conference at the 2012 AIChE Annual Meeting

Pittsburgh, Pennsylvania, USA
28 October - 2 November 2012

ISBN: 978-1-62276-752-6

Printed from e-media with permission by:

Curran Associates, Inc.
57 Morehouse Lane
Red Hook, NY 12571

Some format issues inherent in the e-media version may also appear in this print version.

Printed by Curran Associates, Inc. (2013)

For permission requests, please contact AIChE
at the address below.

AIChE
3 Park Avenue
New York, NY 10016-5991

Phone: (203) 702-7660
Fax: (203) 775-5177

www.aiche.org

Additional copies of this publication are available from:

Curran Associates, Inc.
57 Morehouse Lane
Red Hook, NY 12571 USA
Phone: 845-758-0400
Fax: 845-758-2634
Email: curran@proceedings.com
Web: www.proceedings.com

TABLE OF CONTENTS

(43a) Looking to the Future in Continuous Pharmaceutical Manufacturing

Richard D. Braatz, Chemical Engineering, Massachusetts Institute of Technology, Cambridge, MA

This presentation describes the motivations, challenges, and opportunities in continuous pharmaceutical manufacturing. Some of the motivations are particularly intriguing, such as semicontinuous and continuous-flow unit operations that have the potential to have order-of-magnitude higher controllability and/or are much easier to scale up than batch and semibatch processes. Theoretical, computational, and experimental results are provided as evidence of the potential opportunities of continuous-flow pharmaceutical manufacturing operation. One of the conclusions of reviewing such unit operations is that they can look and operate very different from the batch and semibatch unit operations currently used in the pharmaceutical industry.

Example unit operations are drawn from the published literature, which include both journal papers and patents from various academic and industrial research groups. The presentation will also review and comment on the application of Quality-by-Design methodologies to pharmaceutical manufacturing. Of particular interest is the generation of design spaces, and the pros and cons of different approaches. In particular, it is shown that many of the methods for the construction of design spaces that have been proposed in the literature are useless from a control or operations point of view.

(43b) New Challenges and Opportunities for Pharmaceutical Manufacturing Science

Vincent Vilker, Office of Testing and Research, FDA, Chevy Chase, MD and Mansoor A. Khan, Cder/Ops/Otr/Dpqr, Food and Drug Administratin, Silver Spring, MD

New Challenges and Opportunities for Pharmaceutical Manufacturing Science

Vincent L Vilker and Mansoor A Khan, CDER/FDA

The increasing sophistication of modern drug products is leading to unprecedented challenges and opportunities for pharmaceutical manufacturing. At the same time, scientific and technological advancements in analytical instrumentation, coupled with implementation of the principles embedded in quality by design, enable the improved process monitoring and control that lead to better and more consistent product quality. ICH Q8(R2), the first of three International Conference on Harmonization (ICH) Guidance for Industry that spell out the concepts of Quality by Design, states that critical formulation attributes and process parameters should be identified through an assessment of the extent to which their variation can have impact on the quality of the drug product. In this presentation, examples will be given where modern instrumentation and manufacturing practices are required to maintain the critical formulation attributes of several drug products that have unique physicochemical and/or biological properties. Examples may include properties such as very low solubility, polymorphic solid states, poorly defined macromolecular size distributions, or where there is a tendency to become unstable in the formulated product.

() Recent Advances in the Molecular Modeling Foundation of Systems-Based Pharmaceutics

Constantinos C. Pantelides, Department of Chemical Engineering, Imperial College London,Center for Process Systems Engineering, London, United Kingdom

There are potentially significant benefits from the adoption of a systems-based approach to pharmaceutical science and engineering based on detailed mathematical first-principles models. However, the success of such an approach is predicated on the accurate description of the thermodynamic, kinetic and mechanical properties of the materials involved.

In this presentation, we review some recent advances in the description of the solid and fluid phases arising from work by the Molecular Systems Engineering group at Imperial College London. In particular, we review the state of the art for the *ab initio* prediction of crystal structure and polymorphism, and the prediction of vapour/liquid and liquid/solid phase equilibria using the SAFT-gamma group-contribution based equation of state. Significant progress has been achieved in all these areas over the last few years, but there are also significant challenges that still need to be addressed.

(43d) Experience with Regulatory Issues for Biomedical Devices and Vision for the Use of Advanced Control In Pharmaceutical Operations

Francis J. Doyle III, Chemical Engineering, UCSB, Santa Barbara, CA, Eyal Dassau, UCSB, Santa Barbara, CA and Howard Zisser, Sansum Diabetes Research Institute, Santa Barbara, CA

The landscape for clinical testing of biomedical devices has evolved rapidly over the last 5-10 years. In particular, the use of advanced control algorithms, such as model predictive control, are now being reviewed by the FDA as components of closed-loop devices. In our work with the Artificial Pancreas (AP), we have developed master files and other supporting materials to support IDE (Investigational Device Exemption) studies. In this talk, I will share some of the challenges in developing algorithms for regulatory review, including the role of simulation studies, communication architectures, algorithmic design, and IDE studies. I will extrapolate from these experiences to project the opportunities for advanced control in pharmaceutical operations.

Chemical Engineering & Biopharmaceutics: Linking the Process with the Ultimate Quality Attribute

Ravi M. Shanker

The discovery and development of heterocyclic, small molecule new drug entities is a highly sophisticated, multi-disciplinary, scientifically advanced, collaborative, highly competitive, extremely expensive and multi-year endeavor. The heavily R&D dependent pharmaceutical industry utilizes cutting edge scientific and technological tools and techniques encompassing the broad disciplines of chemistry, biology and medicine. To provide significantly improved therapeutic treatments for various acute and chronic diseases and to be able to meet the requirements and expectations of the regulatory agencies, the patients and the society is a constant challenge for the innovative pharmaceutical industry. The challenges forge deeper dives in science among the wide array of disciplines that constitute the pharmaceutical industry and consequently have provided tremendous opportunities for the advancement of computational and predictive tools within each of these disciplines and sub-disciplines. A key segment of R&D and industrial scale manufacture is pharmaceutical sciences. It is an amalgamation of applied disciplines of chemical engineering, physical and physical organic chemistry, synthetic organic chemistry, process chemistry, analytical chemistry, polymer chemistry, material sciences, crystallization science, drug product design and manufacture and evaluation of performance of drug product in the patients. Pharmaceutical sciences have also embraced the development and use of predictive computational tools among each of its areas of specialization and disciplines to ensure the highest quality of drug product being delivered to the patient. The multi-disciplinary components of pharmaceutical sciences are operationally integrated to provide the highest quality of drug product to meet the regulatory requirements in this highly regulated industry. However, there is a lack of integration of such predictive tools across disciplines to enable the assessment of local perturbations in quality during any segment of work flow on the ultimate quality and performance of the drug product. The ultimate quality attribute of a drug product is it's performance in patients, however, linking that performance to the upstream attributes of drug substance and drug product via engineering control during manufacture with emphasis on the principles of chemical engineering continues to be unrealized and unfulfilled possibility. The vision to provide such integrated approach of existing computational and predictive tools within disciplines and sub-disciplines in pharmaceutical sciences and to provide opportunities for further development of advanced integrated tools as a holistic system is the underlying concept for the creation of Systems Based Pharmaceutics.

Using Models to leverage process understanding; Case studies in O2 sensitivity of a Pd catalyzed CN bond forming reaction and isobutylene release in a acid catalyzed tertbutyl deprotection reaction.

Jeremy M. Merritt[1], Jonas Buser[2], Neil Kallman[1], Mark Pietz[1], Kevin Sullivan[1], Shekhar Viswanath[1], and Jared Fennell[1]

[1]Chemical Product R&D

[2]Analytical Sciences R&D

Eli Lilly and Co.

Abstract:

Oxygen sensitivity was observed in a Pd catalyzed amination reaction employed in the synthesis of an API. Dissolved O2 probes, 31P NMR, and kinetic modeling were used to help elucidate the oxidation mechanism and kinetics, to aid in process understanding and realization of the risks associated with scale-up. Elsewhere in the synthesis, an acid catalyzed tertbutyl deprotection reaction is employed. The largest observed API related impurity formed during the reaction results from re-attack of the tertbutyl group onto another part of the molecule. Several other side reactions were observed including reaction of the tertbutyl with the solvent system (HCl / water / isobutanol) to form tertbutanol, tertbutyl chloride, and isobutyl tertbutyl ether. Due to evolution of gaseous isobutylene during the reaction, there was concern over scale variability due to different gas-liquid mass transfer rates (k_la). A model was developed to predict the reaction rates and equilibria of the main and various side reactions and was used as an argument for scale invariability of the chemistry which was important in defining the impurity control strategy.

(99b) Modeling the Dynamics of the Film Coating Operation for Pharmaceutical Tablets

Salvador García-Muñoz, Process Modeling & Engineering Technology, Pfizer Worldwide Research & Development, Groton, CT, Mariana Moreno, Chemical Engineering, Tec de Monterrey, Monterrey, Mexico, Alejandro Cano, Process Systems Enterprise, Inc., Cedar Knolls, NJ and Simon Leyland, Process Systems Enterprise Inc, Cedar Knolls, NJ

Previous work in understanding the thermodynamics of the film coating operation has focused mainly around the pseudo-steady state reached between the drying air and the cooling effect from the evaporation of the spraying solution [1]. The work presented here aims to mathematically describe the time-varying state of the operation and provide a structured procedure with the optimal experimental approach to estimate the heat and mass transfer coefficients across the system.
The ultimate objective of this work is to deliver a model able to predict the temperature of the tablets and the amount of water in the tablets during (and at the end of) the operation. Such a model will aid the design of the operating cycle for the film coating process if there were constraints on the maximum temperature/water activity that the tablet can be exposed to, or the final amount of water in the tablet core. Such constraints are not uncommon in water sensitive products since this can accelerate degradation mechanism and limit the shelf life of the product.
To obtain a model that is accurate enough (given the uncertainties in the measurements); mass and energy balances were calculated around: drying air, nozzle air, liquids and solids in the solution, tablet cores and equipment. The model is implemented and all parameter estimations done using the gPROMS (Process Systems Enterprise Ltd, London, UK) modeling platform.
A key component to the model is the phenomena occurring at the surface of the tablet; that determines whether the water available in the impacting droplet is absorbed by the tablet or evaporated into the air. This mechanism is modeled as a function the thermodynamic state of the droplet as well as the hygroscopicity of the formulation, mass and surface area of the tablet, the vapour transmission properties of the coating solution and the extent of coating coverage in the tablet. The thermodynamic state of the droplet at the point of impact with the tablet is made dependent on the initial droplet size (calculated using rheological properties, nozzle geometrics, air pressure and the model by Aliseda et al [2]), the chemistry of the coating solution, the thermal conditions in the chamber and the distance traveled by the droplet (gun to bed distance).
The heat capacity of the equipment and the heat transfer from and to the equipment plays an important role to effectively model the energy transfer across the system (which in turn affects the prediction of temperatures). This is due to the fact that equipment becomes an energy sink that can absorb or emit energy in a non-linear way. These equations are established and coupled with the rest of the energy and mass transfer equations.
The model parameters are divided into those that characterize the equipment, the coating solution, the tablets and formulation, the tablet-gas mass and energy transfer and the tablet/liquid mass and energy transfer. Model-guided experiments are carried out in sequence to independently estimate each of the different interactions. The model is finally verified with 21 coating runs in an LDCS 20 coater using tablets of different sizes at multiple thermal conditions and extents of water exposure to the cores. The model is able to provide predictions well inside the experimental accuracy.

References

[1] M.T.am Ende and A.Berchielli, A Thermodynamic Model for Organic and Aqueous Tablet Film Coating, Pharmaceutical Development and Technology, 10 (2005) 47-58.
[2] A.Aliseda, E.J.Hopfinger, J.C.Lasheras, D.Kremer, A.Berchielli, and E.K.Connolly, Atomization of viscous and non-newtonian liquids by a coaxial, high-speed gas jet. Experiments and droplet size modeling, International Journal of Multiphase Flow, 34 (2008) 161-175.

(99c) Modeling of Fluid Bed Granulation Process for Aqueous Systems

Rajan Gupta, Pharmaceutical Commercialization Technology, Merck, Summit, NJ

This presentation discusses the development of a fluid bed granulation model and how it can be applied in the context of QbD. The model uses mass and heat balance to predict a complete profile of bed humidity, product temperature, and exhaust conditions during granulation and drying steps. It also incorporates empirical relations built from actual product data to account for heat losses, tortuosity aspects (particularly important during falling rate period), and sub-saturated conditions. The only inputs required in the model are typical batch parameters (such as spray rate, inlet air temperature / airflow, etc.), thus removing the need to know equipment (e.g., heat transfer coefficients) and formulation related factors (e.g., particle size) in advance.

The finalized model was tested on over 40 historical and new batches, and good agreement was observed between the theoretical predictions and the actual values. The model developed in the current work can be used as a valuable tool in feasibility studies (choosing appropriate parameters as starting points), process scaling-up (lab to commercial), and design space studies.

(99d) Plant-Wide Hybrid Model Predictive Control of a Continuous Pharmaceutical Tablet Manufacturing Process Via Direct Compaction

Ravendra Singh, Marianthi Ierapetritou and Rohit Ramachandran, Department of Chemical and Biochemical Engineering, Rutgers, The State University of New Jersey, Piscataway, NJ

QbD and PAT requires the next generation of pharmaceutical products be manufactured through continuous processing integrated with online monitoring tools coupled with efficient automatic model-based feedback control system to achieve the precise control of process variables so that the predefined end product quality can be achieved consistently.

The direct compaction process considered in this study is highly interactive, nonlinear, and has time delays due to sensor placements, that indicate that a simple feedback regulatory control system (e.g. PI(D)) by itself may not be sufficient to achieve the tight process control as desired by regulatory authorities (e.g. FDA). In previous work, we have demonstrated that under certain conditions, such a control system may not be adequate [1]. Advanced model predictive control (MPC [2]) is the desired control strategy for this complex system especially for systems with multi-variate interactions. As a closed loop optimal control method based on the explicit use of a process model, model predictive control has proven to be a very effective controller design strategy over the last twenty years and has been widely used in process industry such as oil refining and chemical engineering. However, because of different level of complexity involved (e.g. solid handling, irregular flow) in addition to the unavailability of the suitable process model, the design and implementation of MPC in the pharmaceutical manufacturing involving the solid dosage form is still state of the art. In comparison to the regulatory control strategy previously proposed [1], the MPC is more complex and requires the development of higher fidelity models and robust optimization strategies. Depending on the process dynamics either linear MPC (LMPC) that required linear model and less computational power or non-linear (NMPC) MPC that required first principle model and more computational power can be used. However, with the state of the art computational power, the NMPC is not practical for the plant-wide control of complex multi variable system but it could be used efficiently to control the part of a plant which is highly nonlinear. Similarly, LMPC cannot be efficient for the control of the part of the plant which is highly non-linear. Therefore, a hybrid control strategy integrating the regulatory control system, LMPC and NMPC is highly desired for plant-wide control of a process comprising of fast, delayed, linear and non-linear dynamics.

In the work reported here, a hybrid control system integrating PID, LMPC and NMPC for plant-wide control of a continuous tablet manufacturing process has been developed. The continuous manufacturing process involved the state of the arts continuous unit operations including blending, tablet pressing and coating. Based on the analysis of the process dynamics, linearity and non-linearity analysis, the control variables has been placed into three groups: (i) the variables for which the PID controller is sufficient, (ii) the variables for which LMPC is required, (iii) the variables for which NMPC is essential. These variables have been then controlled with the corresponding control strategies. The categorization of these variables ensures the better control performance with optimal resources and computation power. The first principle model as well as the linear model of the continuous tablet manufacturing process has also been developed. The performance of the hybrid control system has been evaluated and compared with the regulatory control strategy. The proposed hybrid strategy is currently being implemented in a continuous pilot plant at Rutgers University.

The objective of this presentation is two-fold: first to highlight the design of hybrid control system for a continuous tablet manufacturing process, and second to demonstrate the performance of the designed hybrid control system *in silico*.

References

[1] Ramachandran, R., Arjunan, J., Chaudhury, A, Ierapetritou, M. (2012). Model-Based Control Loop Performance Assessment of a Continuous Direct Compaction Pharmaceutical Processes. J. Pharm. Innov., 6(3), 249-263.

[2] Glaser, T., Sanders, C. F. W, Wang, F. Y., Cameron, I. T., Lister, J. D., Poon, J. M.-H., Ramachandran, R., Immanuel, C. D., Doyle III, F. J. (2009). Model predictive control of continuous drum granulation. Journal of process control, 19, 615-622.

(99e) PAT for Pharmaceutical Extrusion Monitoring and Supervisory Control

Patrick Wahl[1], Daniel Markl[1], Daniel Treffer[2], Jose C. Menezes[3], Gerold Koscher[1], Eva Roblegg[4] and Johannes G. Khinast[1,2], (1)Research Center Pharmaceutical Engineering GmbH, Graz, Austria, (2)Institute for Process and Particle Engineering, Graz University of Technology, Graz, Austria, (3)Research Center for Pharmaceutical Engineering, Graz, Austria, (4)IPS/Department of Pharmaceutical Technology, University of Graz, Graz, Austria

Introduction

Hot Melt Extrusion (HME) is increasingly attracting interest in pharmaceutical manufacturing, for two main reasons: (1) it allows to produce controlled release formulations by embedding API crystals in polymers, lipids or sugars (Kleinebudde, 2011) and (2) it significantly enhances the bioavailability of APIs with low solubility. This can be achieved with a solid solution, in which the API and the matrix are molecularly dispersed. Estimations predict that up to 90% of future new chemical entities may be classified as poorly soluble. Therefore, enhancing the bioavailability is a significant part for successful drug development. Further dosage forms like transdermal patches and implants among others can be produced by HME as well (Breitenbach, 2002).

HME also has the potential to implement continuous production in a straightforward way. One way is to use a hot-strand cutter as a downstream processing unit. The (hot) extrudate is cut directly at the die face, thereby forming nearly spherical pellets. If necessary, pellets with different APIs or release characteristics can be mixed and further processed by a capsule filling machine or a tablet press. This production line is in development at RCPE.

At present, many Process Analytical Technology (PAT) tools are available for establishing physicochemical product properties, such as the chemical composition. Multivariate data analysis (MVDA) combined with spectroscopic techniques, including near infrared (NIR) and Raman, allows quantitative and non-invasive process monitoring (Koller et al., 2011). Applying multivariate statistical process control (MSPC) reduces real-time process data streams to a convenient control chart. Clearly, process analysis calls for an IT infrastructure that can aggregate real-time process data from multiple unit operations, raw material data, PAT data and equipment status. SIPAT is a commercial PAT software solution that was created to meet the above requirements.

In this study the PAT concept consisted of: (1) In-line, real-time monitoring of extruder parameters and collection of NIR spectra of the extrudate close to the die; (2) Analysis of the spectrum with multivariate data analysis (MVDA) tools; (3) Integration and implementation of this PAT concept in SIPAT.

In this presentation we intend to show the setup of the HME system, Principal Component Analysis (PCA) to reduce the dimensionality of the spectra and a Partial Least Squares (PLS) model to predict the response of the extrusion process (the API content at the die) to step changes of the feed rates.

Experimental Methods

Materials

Paracetamol, donated by G.L. Pharma GmbH, Lannach, Austria, was used as an API (volume mean particle size 139.2 µm). The matrix carrier system was calcium stearate (Werba-Chem GmbH, Vienna, Austria; mean particle size 16.62 µm) (Roblegg et al., 2011). The paracetamol crystals were embedded in CaSt, resulting in a solid dispersion.

Extruder setup

Extrusions were performed on a co-rotating twin-screw extruder (ZSK 18, Coperion GmbH, Stuttgart, Germany) with 18mm screw diameter. The barrel temperatures and the screw setup was chosen in such way that the melt temperature at the die was 130°C. Two twin-screw gravimetric feeders (KT20, K-Tron, Pitman, USA) were filled with a premix. Mixtures with a mass fraction of paracetamol from 0% to 60% can be obtained by adjusting the feed rates of the two feeders. In the experiments the throughput of the extruder was 0.6 kg/h. To monitor, control and store process parameters of the extruder, NIR spectra and results of calculations (e.g. API content prediction) the software SIPAT 3.1.1 (Siemens AG, Brussel, Belgium) was used.

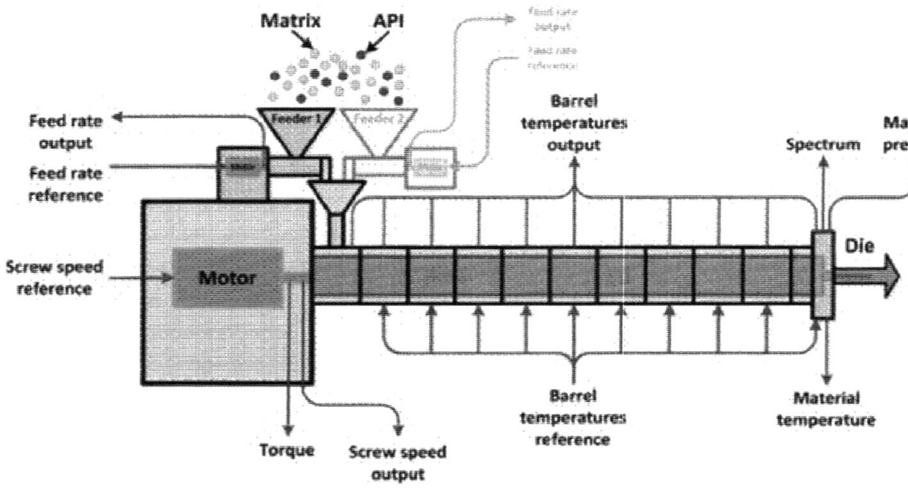

Figure 1: Overview of monitored (green, output) and controlled (red, input) parameters of the extruder and the NIR spectrum, which was taken close to the die.

NIR spectroscopy

NIR spectra were collected in transflexion mode with a process spectrometer SentroPAT FO (Sentronic GmbH, Dresden, Germany). The spectrometer has a spectral range of 1100 nm - 2200 nm and 2 nm resolution. A fibre-optic Dynisco NIR probe was mounted close to the extrusion die, after the screws, for in-line monitoring of the melt. For each measurement 120 spectra were averaged, with an integration time of 0.014s per spectrum. However, significant modifications of the die were necessary to obtain spectra which reflect the current bulk composition of the melt.

Chemometric models

Around 100 spectra of the melt of each step of API content were collected to develop a PCA and a PLS model. Principal component analysis applied to the observed data may be used to classify the main events affecting the process. The PLS model enables the direct prediction of the paracetamol content. The chemometric model was built with the software Simca P+ 12.0 (Umetrics, Umeå, Sweden). Simca-Q 12.0.1 (Umetrics, Umeå, Sweden) is used as an external calculation engine in SIPAT, which has the advantage of directly integrate the models developed in Simca P+ in SIPAT.

Results

Selecting a good NIR measurement position

Ensuring a good sample presentation to the Dynisco NIR probe is crucial for correct API predictions. The main reason for improper sample presentation is a too slow exchange of melt in the field of view of the NIR probe. First, if the probe is not perfectly in line with the barrel surface, melt will fill the dead volume once. The shear forces are not high enough to exchange the material. This is likely for measurements above the screws. A second possibility is a too large volume of melt directly before exiting the die. Here a funnel flow might build up, which causes slow melt exchange at the barrel walls, where the probe is situated. Although NIR penetrates a transparent melt, in the mentioned cases the measurement could be biased. Thus, a special probe set-up was designed and tested to represent the actual melt composition.

Monitoring the process

The feeding dynamics are the most important factor for content uniformity of the melt, especially for pharmaceutical powders, which are often not free flowing. Despite the mixing elements of the screw, the extruder has only a limited capability to compensate lower frequency feeding variations by its backmixing ability.

Monitoring the process in the reduced PC space is a straightforward way of implementing real-time process analysis. Adjusting the feed rates of the feeders yielded different API concentrations in the extrudate that appeared in the score plot as clusters of observations.

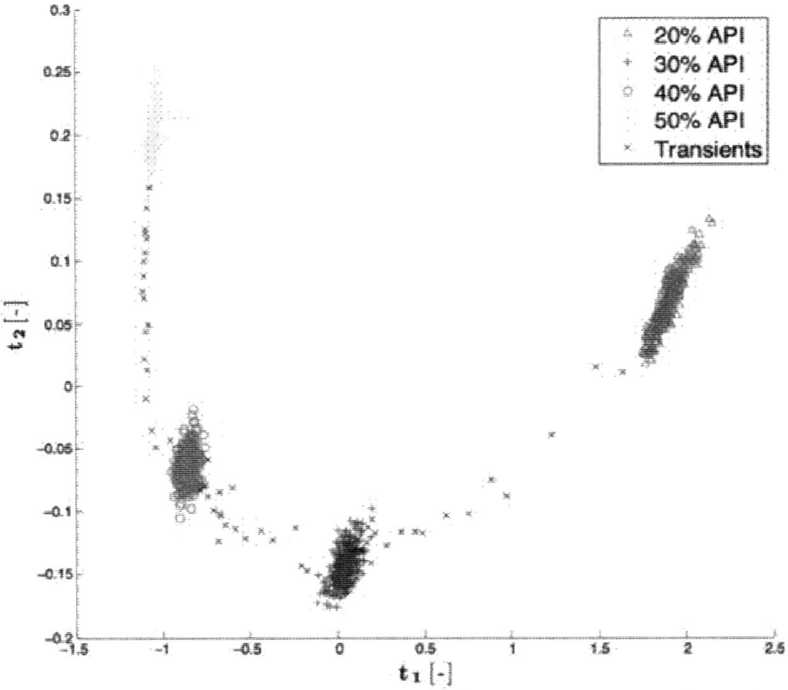

Figure 2: Different API concentrations (20%, 30%, 40% and 50%) result in the clustering of samples in the score plot (t_1 versus t_2).

Moreover, the prediction of the API content by the PLS model enables another way of monitoring the process in a convenient control chart.

Using Simca P+ models or models defined in MATLAB, the observations can be processed and visualized in SIPAT in real-time. The integration of the PCA and PLS models facilitates different ways to detect in real-time abnormalities in the process.

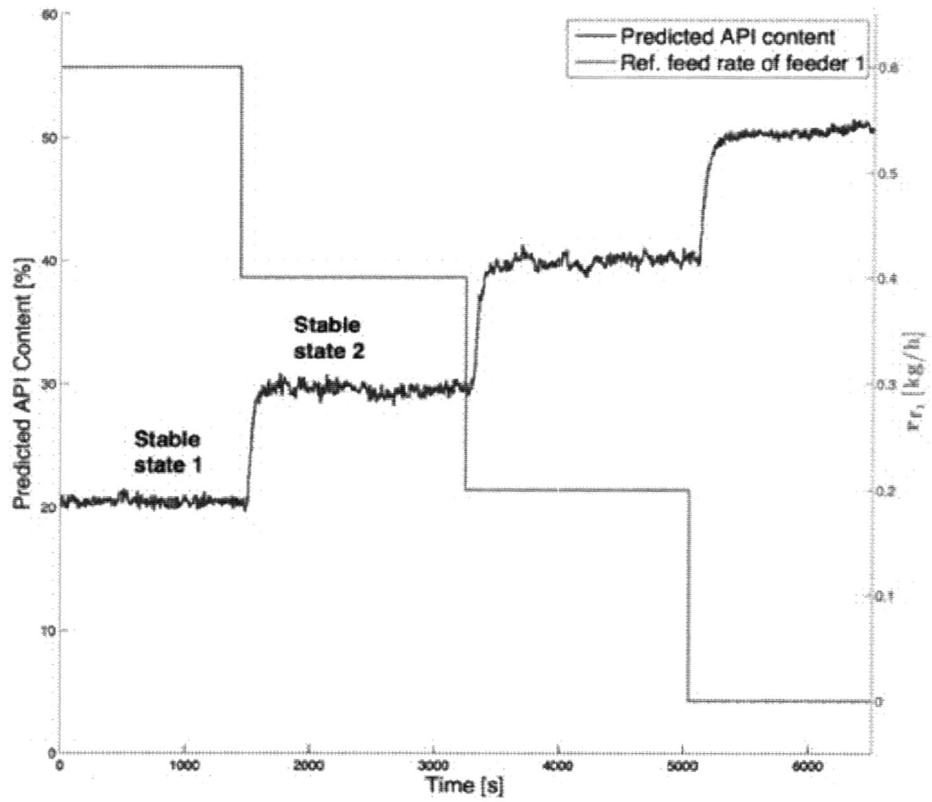

Figure 3: Predicted paracetamol content (blue). Variations induced by the feeders are noticeable. For the chosen screw configuration the time delay between feeder settings (magenta) and predicted API content at the extrusion die is 55s.

Conclusion

The presented hot melt extrusion system with an integrated PAT concept is capable of monitoring the critical process parameters. Special attention was paid to ensure a good NIR measurement position. However, the MVDA models that are developed in this process can be applied to only one formulation. In a typical campaign-driven manufacturing environment (which even continuous manufacturing of the future will be) formulations and/or products may change weekly or monthly. Automatic model development can be performed based on the setup including the HME process, SIPAT and Simca-Q. Using several models at the same time opens up the opportunity to validate different PCA or PLS models in real-time.

References

[1] Breitenbach J (2002) Melt extrusion: from process to drug delivery technology. *Eur. J. Pharm. Biopharm.* **54**:107-117.

[2] Kleinebudde P (2011) Pharmaceutical Product Design: Tailored Dissolution of Drugs by Different Extrusion Techniques. *Chem-Ing-Tech.* **83**:589-597.

[3] Roblegg E, Jäger E, Hodzic A, Koscher G, Mohr S, Zimmer A and Khinast J (2011) Development of sustained-release lipophilic calcium stearate pellets via hot melt extrusion. *Eur. J. Pharm. Biopharm.* **79**:635-645.

[4] Koller D M, Posch A, Hörl G, Voura C, Radl S, Urbanetz N, et al. (2011) Continuous quantitative monitoring of powder mixing dynamics by near-infrared spectroscopy. *Powder Technology*. **205:** 87-96

2012 AIChE Annual Meeting QbD Topical Conference
"Model Based QbD in Drug Substance or Drug Product Development" Session
Pittsburgh, PA, Oct. 28-Nov. 2, 2012
Paper 99f

Scientific Validation of THz Spectroscopic Measurement and its Pharmaceutical Process Analytical Technology (PAT) Applications: Case Studies of First Principles Modeling Approach

Huiquan Wu* and Mansoor Khan

Division of Product Quality Research (DPQR, HFD-940)
OTR/OPS/CDER/FDA
FDA White Oak Research Center
10903 New Hampshire Ave
Silver Spring, MD 20993

Abstract: As an emerging technology, THz spectroscopy has gained increasing attention in the pharmaceutical area during the last decade. This attention is due to the fact that (1) it provides a promising alternative approach for in-depth understanding of both intermolecular interaction among pharmaceutical molecules and pharmaceutical product quality attributes; (2) it provides a promising alternative approach for enhanced process understanding of certain pharmaceutical manufacturing processes; and (3) the FDA pharmaceutical quality initiatives, most noticeably, the Process Analytical Technology (PAT) initiative. In this work, the current status and progress made so far on using THz spectroscopy for pharmaceutical development and pharmaceutical PAT applications, including FDA-NIST collaboration on THz spectroscopy pharmaceutical characterization and PAT applications, are discussed. In the spirit of demonstrating the utility of first principles modeling approach for addressing model validation challenge and reducing unnecessary model validation "burden" for facilitating THz pharmaceutical PAT applications, two scientific case studies are created and discussed. Furthermore, other technical challenges and opportunities associated with adapting THz spectroscopy as a pharmaceutical PAT tool are highlighted.

Keywords: Terahertz Spectroscopy, Pharmaceutical Product Quality Control, Process Analytical Technology (PAT), Chemometrics, Pharmaceutical Manufacturing Process, First Principal Modeling, Process Understanding.

Disclaimer: The views and opinions expressed in this paper are only of authors, and do not necessarily reflect the view or policies of the FDA.

*Corresponding author: Phone 301-796-0022; Fax: 301-796-9816; Email: Huiquan.wu@fda.hhs.gov

1. Introduction

The FDA recently released a strategic plan about advancing regulatory science at FDA [1]. One of the eight science priority areas identified in that strategic plan is "support new approaches to improve product manufacturing and quality". For implementation strategy, "FDA will support the application of novel technologies to product development and innovative analytical approaches to improve product manufacturing and quality…" by "enabling development and evaluation of novel and improved manufacturing methods", including "prompting two state-of-the-art manufacturing strategies-Process Analytical Technology, and Quality-by-Design approaches—for impact on manufacturers' ability to maintain consistent quality." Spectroscopic techniques and material characterization techniques have been playing a critical role for gaining pharmaceutical product and process understanding, ensuring pharmaceutical product quality [2] under the umbrella of the FDA's Pharmaceutical Quality Initiative for the 21st century [3], including Process Analytical Technology (PAT) [4] and Quality-by-Design (QbD) [5] approaches. It has been well recognized that process validation in the PAT domain could be challenging, as discussed in the FDA's new draft guidance on process validation [6].

Terahertz (THz) spectroscopy has been widely used to study the spectroscopic characteristics of a variety of materials in the spectral region spanning from 0.1 to 10 THz. The majority of crystalline solids have distinct and characteristic lattice vibrations within the THz frequency range, making THz spectroscopy a powerful analytical tool for probing the intermolecular interactions of molecular crystals. As a matter of fact, THz spectroscopy technique shows promising for the analysis of pharmaceutical products throughout the drug development process [7-8]. Most noticeably, THz spectroscopy as an emerging technology has been gained increasing attention in the area of pharmaceutical product and process understanding over the last several years. In this work, current status and progress on using THz spectroscopy for pharmaceutical product quality control and pharmaceutical Process Analytical Technology (PAT) [2] applications were reviewed. Two case studies of using first-principle approach to validate THz spectroscopic measurements were discussed. Challenges and opportunities associated with adapting THz spectroscopy as a pharmaceutical PAT tool were highlighted.

2. THz spectroscopy for pharmaceutical product characterization and pharmaceutical development

In this regard, THz spectroscopy has been applied to certain important areas, which includes but is non limited to:
(1) Crystalline pharmaceuticals, polymorph screening, and solid-state assessment [9-16];
(2) Assessing weak interaction (inter-molecular vibrations) and intra-molecular interaction [17-20];
(3) Mapping chemical compositions and tablet density [21-22];
(4) QA and QC for coating and controlled release dosage form manufacturing, dissolution performance, tablet thickness
Details could be found in the literatures cited.

3. THz spectroscopy for pharmaceutical PAT applications

According to the FDA's PAT Guidance for the Industry [2], PAT is a system for designing, analyzing, and controlling manufacturing through timely measurements (i.e., during processing) of critical quality and performance attributes of raw and in-process materials and processes, with the goal of ensuring final product quality. An ultimate goal of PAT is to enhance understanding and control the manufacturing process. Consequently, the tools and principles described in this guidance should be used for gaining process understanding and can also be used to meet the regulatory requirements for validating and controlling the manufacturing process. Since the FDA launched its PAT initiative in 2002, broad interests have been stimulated for the pharmaceutical community to adopt many analytical and characterization techniques for at-line, on-line, and in-line pharmaceutical PAT applications. Among these innovative spectroscopic pharmaceutical applications, THz spectroscopy has found some important yet promising applications due to its unique advantages. Some of the THz spectroscopy pharmaceutical PAT-oriented applications, most for small molecule drugs and several for biopharmaceuticals, are discussed below.

3.1. Film coating process of pharmaceutical tablets

Tablet coating thickness is a critical quality attribute (CQA) for modified (extended) release (MR or ER) or controlled release (CR) tablet dosage forms, as it directly dictating the dissolution rate [23] and dissolution performance of the MR/ER/CR tablet dosage form, thus linking to the pharmacological effect (AUC, t_{max}, etc.) and bioavailability. Traditionally, the tablet coating thickness is calculated based on the tablet weight gain obtained during the coating process. Direct measurement methods, such as scanning electron microscopy (SEM), need to cut the tablet to get the photographs of cut-through cross-section, thus are destructive and prone to error. In recent years, spectroscopy techniques have provided fast and non-destructive alternate approaches to evaluate the tablet coating and coat performances.

3.1.1 THz Pulsed Imaging for nondestructive analysis of tablet coating thickness

Few nondestructive techniques are capable of examining tablet coatings. THz Pulsed Imaging (TPI) is a fast (a waveform can be generated in less then 20 ms) and nondestructive technique. It was reported that TPI can distinguish two brands of over-the-counter ibuprofen tablets [24]. Good agreement was obtained between THz maps obtained via TPI and photographs of cutting-through sections of tablets. It is nondestructive, noninvasive, safe, fast, and provides information on factors influence bioavailability. Technically it can probe any position on a tablet in less than 20 ms, making it possible for rapid online assessment. Furthermore, the variation of internal coating layers can be observed which would not be readily apparent from separated and isolated point measurements. However, lateral resolution is limited by diffraction of the THz focus to about 150 μm at 3 THz, whereas the axial resolution is limited by the THz pulse duration (which is <200 fs) to about 30 μm. Therefore, more development work is to be done before TPI can provide reliable results on thin coating layers, especially for thickness <30 μm.

3.1.2. THz Pulsed Imaging (TPI) for assessing sustained-release tablet film coating quality including coating uniformity, dissolution performance; Critical assessment

Using TPI, Ho et al [25-27] conducted a thorough study to examine sustained-release tablet film coating quality including coating uniformity and how they impact dissolution performance of the coated tablets. The relationship between THz parameters (THz electric field peak strength (TEFPS) and coating layer thickness) and product performance (the Mean dissolution time, MDT) using SR tablets obtained during a scale-up process was examined. It was shown that weight gain (the traditional non-destructive indicator for coating quality in the dry-state) failed to reflect the functional performance of the product (with R^2 value of 0.09). However, THz parameters could be correlated with the actual product performance (with R^2 value around 0.84), thus showed potential to predict the dissolution performance of SR tablets. The proof-of-concept was proved. However, as shown in their Figures 3A-C, the R^2 values were based on two clusters of data points which are located at two ends of the regression lines. This situation will negatively impact the prediction power of the regression equation. Ideally, those data points should be evenly distributed along the regression line. A more systematic study to provide sufficient statistical power to further understand the extent of the relationship between the dry-state THz parameters and product performance is warranted. They also used TPI to determine film coating thickness and TEFPS for analysis of SR coated pellets [26]. Their results suggested that drug release for pellets studies was not predominately governed by drug diffusion through the polymeric film coating, but probably to a large extent by drug solubility. TPI's capability to detect non-homogeneities in the drug layer and polymeric film coating was demonstrated. Most recently, Ho et al. [27] using TPI to investigate dissolution performance critical areas on coated tablets, i.e., formation and location of film coating "weak spots" or film coating defects. TPI analysis showed that for the film coated tablets studied, the tablet central band exhibited the thinnest film coating, lowest coating density, and highest surface roughness, thus was the performance limiting area of the film coating. Dissolution tests confirmed these TPI observations: the film coating on the tablet central band has a faster MDT of 7.4 hr in comparison to 10.4 hr for the convex top/bottom surface. Therefore, TPI has potential to be adopted as a PAT tool to probe film coating defects during coating development and to evaluate the effect on subsequent dissolution performance.

3.1.3. Comparison of various non-destructive methods for quantifying tablet coating thickness

Zeitler and Gladden [28] reviewed *in-vitro* tomography and non-destructive imaging at depth of pharmaceutical solid dosage form. Four most promising techniques, X-ray computed microtomography, magnetic resonance imaging, THz imaging, and optical coherence tomography, were discussed to introduce the concept of tomography in a pharmaceutical setting. Their operating principles as well as their specific strengths and limitations were discussed. A recent study [29] compared non-destructive methods for quantifying tablet coatings thickness, including optical microscopy, micrometer, X-ray fluorescence (XRF), Raman, and Near-infrared (NIR) spectroscopy. Their conclusion was, direct thickness measurements were time-consuming but provided assured coat thickness data. On the other hand, XRF, Raman, and NIR spectroscopy methods were viable alternatives to provide complementary information for the tablet coatings studied. As discussed previously, THz spectroscopy also provides an important alternative to quantify the coating thickness.

3.2. Biologic molecules and biopharmaceutical applications, medical imaging modality

Johnson et al. [30] reported their study on simultaneous recording the mid-IR and far-IR (THz) spectra of the sporulated form of five *Bacillus* bacteria. Their results indicated that any room-temperature THz absorption features due to spores are at least 28 times weaker (based on p-p noise) than the corresponding mid-IR amide I band, thus limited the utility of the THz domain for the room temperature detection of spores. Tedy's group [31] studied THz absorption spectra of the similarly structured amino acids *L*-serine and *L*-cysteine in the solid phase at 77 and 298K. The THz spectra were compared to isolated molecule and solid-state infrared vibrational spectral calculations using force field and density functional theory. These molecules differ only by a single atom but the low frequency spectra exhibit very different absorption features, which can be attributed to changes of the internal atomic masses and hydrogen-bonding character of the O-H vs. S-H groups within the solid phase unit cell. These differences lead to different intermolecular interactions.

Brandt et al. [32] studied THz time-domain and Raman spectroscopy of the sulfur-containing cystein-based peptide dimmers in the region of low-frequency infrared vibrations at room temperature. The differences between Raman and THz absorption spectra can be attributed to different intermolecular interactions, and thus, the lowest frequencies reflect different lattice vibrations. Therefore, the low-frequency Raman and THz absorption lines reflect both inter- and intra-molecular motions. Typically, low-frequency vibrational modes of biological molecule consist of both intramolecular modes and intermolecular modes. The former is dependent on the molecule as a whole; the latter arises from hydrogen-bonding interactions and van der Waals forces. Therefore, vibrational modes contain information about confirmation dynamics of biological molecules and can be used for identification, as being used for small molecules discussed previously. However, quite often it is challenging to distinguish the broad and overlapping features generated by conventional FTIR spectroscopy and THz time-domain spectroscopy (THz-TDS). Laman et al [33] reported their development of a high-resolution waveguide THZ spectroscopy of biological molecules, which resulting sharper THz features. This technique has been successfully demonstrated with seven small biological molecules, thymine, deoxycytidine, adenosine, D-glucose, tryptophan, glycine, and L-alanine, which shows the possibilities and promise for future studying vibrational modes of large biological molecules. George et al. [34] reported THz time-domain spectroscopy of bovine serum albumin in aqueous solution using Zeonor microfluidic channels. The measured results agree well with those previously reported using high-power THz free-electron lasers. Their work demonstrated the feasibility of performing high sensitivity THz spectroscopy of biomolecules in aqueous solutions with detectable molecular quantities as small as 10 picomoles using microfluidic devices.

Liu et al [35] described sensing minute changes on biological cell monolayers (bovine lung microvessel endothelial cells in response to vascular endothedial growth factor (VEGF)) with THz differential time-domain spectroscopy (THz-DYDS). These changes were reflected by alterations in THz wave attenuations and THz dielectric properties of the treated cells. They revealed that the VEGF-induced THZ attenuations of cell monolayers correlated well with changes in transendothelial resistance measured by electrical cell-substrate impedance sensing (ECIS). However, the morphological differences causing these changes were not observed with standard optical phase contrast microscopy. Thus, THz-DTDS could be a complementary technique for an optical microscopy or ECIS to measure properties of live cells, including cell growth, cell volume, and morphological change over time. Its limitation is that cell samples need to be extracted from the medium for free-space measurements, in order to avoid bulk water absorption.

A thorough review by Pickwell-MacPherson and Wallace [36] discussed the progress on THz Pulsed Imaging (TPI) in its application to biomedical problems with the possibility of developing it into viable medical imaging modality in the future. Several challenges are to be tackled if THz imaging is to become an established medical imaging modality: (1) The origin of image contrast needs to be better understood. Presumably, the changes in optical properties of the tissues observed due to increased water content and structural changes due to the presence of disease are responsible for the contrast. The question is how can we distinguish them from benign conditions? (2) The technology needs to improve such that *in vivo* imaging can be easily performed on a variety of sites within the body and not just on the skin. Given the low penetration depth, THz imaging during surgery or via an endoscope could be possibilities.

3.3. Chemometrics aspects

Like any other spectroscopic technique, for successful development of THz spectroscopy pharmaceutical PAT applications, chemometrics is an enabling tool for extracting critical process and product information from THz data to build up a model. Depending on which type of model is established, such a chemometrics model can be used for different purposes, such as process monitoring and process supervision, quantification, process control, etc. To establish a good model based on THz spectral data for pharmaceutical PAT applications, it is important to recognize and address the following aspects during the life cycle of THz PAT model: (1) the impact of THz data quality and spectral data preprocessing algorithms on predictive power and robustness of the model; (2) the role of Design of Experiments (DOE) in PAT modeling; and (3) model validation, maintenance, and update.

3.3.1. Impact of THz data quality and spectral data preprocessing algorithms on predictive power and robustness of the THz PAT model

Based on the discussion in previous subsections, it is recognized that THz absorption spectral values of pharmaceutical materials and pharmaceutical dosage form are a complex function of a series variables, such as (1) properties, compositions, and structure of the pharmaceutical materials and pharmaceutical dosage form; (2) the measurement conditions; and (3) THz instrumental variables, etc. Our research results [37] demonstrated the effects of compositional and instrumental variables on THz spectral data quality for THz pharmaceutical PAT application. In addition, the influences of spectral data preprocessing algorithms on the predictive power and robustness of the multivariate statistical models have been recognized [23, 38-41]. It is well known that spectral data are information rich, often capturing both chemical and physical information of the substance. It is always challenging to separate the chemical information and physical information embedded with spectral data. If use right, spectral data preprocessing algorithms could be useful to filter certain irrelevant information in the data, thus allows establishing more realistic correlation between the independent variables and responsible variables. As an integral part of chemometrics modeling, the objective of the pre-processing is to remove physical phenomena in the spectra so as to improve the subsequent multivariate regression, classification model or exploratory analysis. There are two types of pre-processing methods which are most frequently used: scatter-correction methods and spectral derivatives. Multiplicative Scatter Correction (MSC), de-trending, Standard Normal Variate (SNV), and normalization are belong to the first group. Savitzky-Golay (S-G) polynomial derivative filter and Norris-Williams (NW) derivative are representatives of the 2nd group. Those data preprocessing algorithms are available in several commercial software packages specialized in handling spectral data for multivariate data analysis and chemometrics modeling.

3.3.2. Role of Design of Experiments (DOE) in PAT modeling

One essential element for pharmaceutical PAT is process understanding. According to the FDA's PAT Guidance, "A process is generally considered well understood when (1) all critical sources of variability are identified and explained; (2) variability is managed by the process; and, (3) product quality attributes can be accurately and reliably predicted over the design space established for materials used, process parameters, manufacturing, environmental, and other conditions. The ability to predict reflects a high degree of process understanding." Design of Experiments (DOE) provides a rationale approach to identify the variability and possible interactions. Integration DOE with PAT real time process monitoring could provide a powerful tool to understand the variability occurred with the process, and more importantly, to reveal how the process variability are translated into product variability. Most importantly, such integration is essential to achieve a high degree of process understanding due to the facts (1) the variability is examined systematically via DOE; and (2) the process information is acquired over the entire time course of the process via PAT real time process monitoring, which allows us to evaluate the process kinetics and process dynamics in a detailed fashion. Therefore, integration DOE and PAT real time process monitoring is essential to PAT modeling, product and process understanding, and design space development. Several FDA internal research projects have discussed those aspects in details [42-47].

3.3.3. Model validation, maintenance, and update

Model validation, maintenance, and update are important aspects for any process models associated with pharmaceutical PAT applications, including THz pharmaceutical PAT applications, if the model developed is to be used for quantification or process control. To the best of our knowledge, currently there is no standard or regulatory guidance that can be specifically address all of issued associated with THz pharmaceutical PAT applications in the aspect of model validation, maintenance, and update. However, relevant references including ASTM standards and EMEA NIR Guideline [48-52] may be adapted to address those issues, at least providing a good starting point.

In author's scientific opinion, the validation requirements for different types of models are different. For first-principles models and mechanistic models, we do need to examine the conditions and assumptions made during the modeling exercise to make sure those conditions and assumptions are still valid during the applications of those first-principles models and mechanistic models. For chemometric models or data-driven empirical models, a significant validation effort is mandatory, including but is not limited to cross-validation as the first safe guard, external validation via independent data set that has not been used previously for model calibration exercise. For this chemometric model, model maintenance and update are important steps to ensure the accuracy, representativeness, and robustness of the calibration model established during the pharmaceutical PAT application, in the event unaccounted sources of variability occurs.

4. Case Studies on First principle modeling approach to validate THz Spectroscopic Measurement

4.1. *In situ* monitoring technique for examining solvent diffusion in polymers: first-principles modeling case study

The ingress of liquid solvent molecules into solid polymers is of importance to certain applications, such as coating of pharmaceutical tablets (for controlled release/modified release dosage form design) and pharmaceutical devices where solvent transport is desirable, and packaging where solvent transport causes degradation and contamination of the polymer. Therefore, accurate measurement of the liquid solvent ingress position within the polymer film is critical for purposes such as validation of the models of the process, improved understanding of the liquid penetration process, and potentially the provision of on-line or at-line measurement during processing or storage as advocated by the FDA's PAT Guidance. Obradovic et al [53] reported a novel application of THz time-domain reflection spectroscopy to measure the penetrant front position of acetone in various polymers (polycarbonate and PVC) *in situ*, and thus as a function of time and temperature. In their work, the reflection of a short THz pulse off a penetrant front was observed. The data of depth of solvent penetration derived from *in situ* THz time-domain reflection spectra were then validated by detailed comparison with Magnetic Resonance Imaging (MRI) measurements on the same systems. The schematic of the THz reflection measurement is shown in Figure 1. The first-principles which govern the measurement process are expressed mathematically below.

The path length differences in polymer x_2 and in air x_1 can be written as:

$$x_2 = \frac{2d_a}{\cos\beta} \tag{1}$$

$$x_1 = 2d_a \tan\beta \sin\alpha \tag{2}$$

Therefore, the overall path length difference Δx is:

$$\Delta x = x_2 - x_1 = \frac{2d_a}{\cos\beta}(1 - \sin\beta\sin\alpha) \tag{3}$$

By substituting Shell's law [54], $\frac{\sin\alpha}{\sin\beta} = \frac{n_2}{n_1}$, we have:

$$\Delta x = \frac{2d_a}{\cos\beta}(1 - \frac{n_1}{n_2}\sin^2\alpha) \qquad (4)$$

The time delay Δt at the detector introduced by the additional path length Δx is given by:

$$\Delta t = t_2 - t_1 = \frac{x_2 n_2}{c} - \frac{x_1 n_1}{c} = \frac{2d_a(t)}{c\cos\beta}(n_2 - n_1 \sin\beta\sin\alpha) \qquad (5)$$

Clearly, the THz reflection measurement results were expressed by first-principles models. Furthermore, good agreement with respect to the actone's penetration front position was obtained for both polycarbonate and PVC, when compared with corresponding MRI data. Therefore, the THz reflection measurement was validated by independent MRI measurement. In author's opinion, this work provides an excellent scientific case study which demonstrated how first-principles models play a critical role in the PAT and QbD implementation.

4.2. *Ex-situ* monitoring technique for examining moisture content in test disk: first-principles modeling case study

Previously [55], we used THz spectroscopy to identify water peak and derive the disk drying process kinetics. Pharmaceutical tablet (Theophylline as the drug with several excipients) was crushed into fine powder, then mixed with polyethylene. The mixture was then compressed into disks of thickness 2 mm in a 13 mm die at approximately 200 psi pressure. All disks were made under lab humidity conditions (relative humidity (RH)= 40 ±1 %). All disks especially the subsurface portions would contain a small amount of water from either moisture absorbed by components (Hancock and Shamblin, 1998) [56] or residual water in the tablets, from the mass balance point of view. Consequently, when the disk is placed into the spectrometer sample compartment for dried-air (RH=0%) purging, any subsurface water would diffuse out and evaporate into the gas phase in the compartment, and then quickly removed by the dried-air purging process, due to the gradient of water levels across the subsurface of disk and the gas phase according to the inter-phase mass-transfer theory [57]. THz measurement was done for various disk drying time points. Good agreement was obtained between the THz experimental results of disk drying, previously identified THz peaks for water, and theoretical mass-transfer analysis of the disk drying process, as discussed below.

Previously [58-61] (Rahman and Stillinger, 1971; Hasted et al, 1985; Thrane et al, 1995; Castner and Chang, 1995) two absorption bands centered at 50 cm^{-1} and 200 cm^{-1} for liquid water were identified in far-infrared measurements. The band at 200 cm^{-1} was broad and corresponding to the stretching mode of the O—H...O unit of the water molecules. The exact position and width could vary slightly, depending on the chemical and physical environment. Our study shows that there are two bands with peak maxima located around 54 cm^{-1} and 210 cm^{-1}on the tablet's corrected THz absorption spectra, which obviously are not associated with characteristic peaks of excipients in the tablet formulations. Meanwhile, two prominent peaks for the theophylline component around 52.1 cm^{-1} and 210 cm^{-1} were observed. Given that theophylline is stable under our experimental conditions, presumably we could assume these two prominent theophylline bands are superimposed on the water bands in their vicinity around 50 cm^{-1} and 200 cm^{-1}. The following derived drying process kinetics analysis further supported this rational.

According to Beer's law, the spectral absorbance (A) is proportional to the concentration (c) of species being detected:

$$A = \varepsilon \cdot l \cdot c \qquad (6)$$

Where ε is the molar absorbiptivity, l is the light path length. As a first appropriation, both ε and l can be assumed constants. Thus, the absorbance difference at the peak maxima can be related to the mass change associated with the disk drying process over time. Furthermore, the absorbance difference divided by the drying time difference can be a measure of the drying rate, if we make a reasonable assumption that the disk surface area remains unchanged during this relative short period of drying time. Figure 2(a)-(d) illustrates the relationship between this derived disk drying rate vs. disk drying time. For the bands centered around 54 cm^{-1} and 210 cm^{-1}, most WG tablets and DC tablets exhibit a distinct two-stage drying process: (1) at the 1st stage, the absolute value of the drying rate decreases very quickly when t_{disk} increased from 5 minutes to 30 minutes; and (2) at the 2nd stage, a nearly constant drying rate (almost zero) was exhibited when t_{disk} increased further to 60 minutes. This two-stage drying model appears to qualitatively agree with the typical drying curve of materials at constant drying conditions, and could be well explained by the interphase mass-transfer theory [57]. Therefore, the bands located at 54 cm^{-1} and 210 cm^{-1} are interpreted as combinations of THz absorption features of theophylline component and moisture. Furthermore, the absorbance difference observed over disk drying time was primarily attributed to residual water associated with the disk due to the high sensitivity of THz spectroscopy to water, although the amount of residual water does not appear to be significant.

This work provides a scientific case study which illustrated how first-principles models (in this case mass-transfer and heat-transfer) were applied to analyze THz test disk drying data to derive disk drying kinetics. The benefits of this approach lie in: (1) identification characteristic THz peaks for water in THz absorption spectra; (2) enabling optimal THz testing conditions to ensure THz data quality for THz pharmaceutical PAT application.

5. Challenges and Opportunities for adopting THz spectroscopy for a pharmaceutical PAT tool

From the public literatures, so far the majority of THz spectroscopy pharmaceutical PAT applications have been focused on real time process monitoring of tablet coating processes. Proof-of-concept has been proved. Preliminary success has been achieved for relative thick coating thickness. However, limited sensitivity to low coating thickness has been a significant challenge for the applicability of this development. On the other hand, from quantitative modeling perspective, a reference method which can directly measure coating thickness in an accurate and reliable manner is much needed. Destructive method (cutting-trough the film) is still most dominant method of directly measuring coating thickness at the moment, in spite of its possibility of introducing additional variability and error. As discussed above, TPI provides an alternate method to detect defects in the film, which might be linked to the dissolution performance critical areas, if the linkage could be established and confirmed by direct and microscopic site evidence.

The ultimate goal for THz spectroscopy pharmaceutical PAT application is to gain in-depth pharmaceutical process understanding, establish a process model, and hopefully use this model to control the pharmaceutical process. To achieve this ultimate goal, such a model must be representative (to the situations examined) and robust (to certain level of noises and/or process disturbance). As most of the works published so far are at laboratory scale, significant challenges and opportunities are coexisting when move the laboratory scale work at R&D to pilot scale and manufacturing scale, as discussed previously [62] and highlighted in the FDA's pharmaceutical quality initiatives in the 21st century [63-64].

6. Concluding Remarks

It can be easily appreciated that THz spectroscopy has found interesting and promising applications in the following areas. (1) pharmaceutical product characterization and pharmaceutical product quality control; (2) pharmaceutical PAT, especially real-time process monitoring of certain pharmaceutical manufacturing processes, etc. Due to its unique capability of probing the intermolecular interaction and intra-molecular interactions, THz spectroscopy provides a novel approach to study pharmaceutical polymorphism, solid-state properties, crystallinity, film coating process, and even biological and biopharmaceutical applications, etc. While direct validation of THz spectroscopy measurement result has been a significant challenge, first-principles models can provide a reliable approach to validate THz spectroscopic measurement, as demonstrated in the two case studies created in this work. Both scientific and regulatory challenges in adopting THz spectroscopy for pharmaceutical PAT applications were discussed in the context of THz measurement science and PAT chemometrics modeling.

Acknowledgement

The author greatly acknowledges scientific discussions on THz technology with Dr. Edwin Heilweil at the National Institute of Science and Technology (NIST), Gaithersburg, Maryland, USA. The authors gratefully acknowledge the original FDA PAT Policy Team, the FDA PAT Review and Inspection Team, the FDA PAT Research Team, and FDA CDER Senior Management Team. FDA CDER Regulatory Science and Review (RSR) Funding Support (RSR-04-16) and Interagency Agreement between the FDA and NIST are greatly acknowledged. Great contributions from the THz field from various technology vendors and research groups are appreciated. The scientific discussion on validation of new technology with Dr. Vincent Vilker, Director, Office of Testing and Research, CDER, FDA, is acknowledged.

References

[1] FDA, August 2011. A Strategic Plan: Advancing Regulatory Science at FDA. Available at: http://www.fda.gov/downloads/ScienceResearch/SpecialTopics/RegulatoryScience/UCM268225.pdf (accessed on 09/24/2012)

[2] J. Woodcock, 2004. The Concept of Pharmaceutical Quality, *American Pharmaceutical Review*, **7** (6), pg. 10-15.

[3] FDA, 2004. Pharmaceutical CGMPs for the 21st Century: A Risk-Based Approach. Final Report. Available at:

http://www.fda.gov/downloads/Drugs/DevelopmentApprovalProcess/Manufacturing/QuestionsandAnswersonCurrentGoodManufacturingPracticescGMPforDrugs/UCM176374.pdf (accessed on 09/24/2012)
[4] FDA, 2004. "Guidance for Industry. PAT—A Framework for Innovative Pharmaceutical Development, Manufacturing, and Quality Assurance". Available at: http://www.fda.gov/downloads/Drugs/GuidanceComplianceRegulatoryInformation/Guidances/UCM070305.pdf (accessed on 09/24/2012)
[5] FDA/ICH, 2009. Guidance for Industry, Q8(R2) Pharmaceutical Development. available at: http://www.fda.gov/downloads/Drugs/GuidanceComplianceRegulatoryInformation/Guidances/UCM073507.pdf (accessed on 09/24/2012)
[6] FDA, 2011. "Guidance for Industry. Process Validation: General Principles and Practices. Rev 1". Available at: http://www.fda.gov/downloads/Drugs/GuidanceComplianceRegulatoryInformation/Guidances/UCM070336.pdf (accessed on 09/24/2012)
[7] J. Aaltonen, K.C. Gordon, C.J. Strachan, T. Rades, 2008. Perspectives in the use of spectroscopy to characterise pharmaceutical solids. International Journal of Pharmaceutics 364(2008)159-169.
[8] P.F. Taday, 2004. Applications of terahertz spectroscopy to pharmaceutical sciences. Trans. R. Soc. Lond. A (2004) 362: 351-364.
[9] P.F. Taday, I.V. Bradley, D.D. Arnone, M. Pepper, 2003. Using Terahertz Pulse spectroscopy to Study the Crystalline Structure of a Drug: A Case Study of the Polymorphs of Ranitidine Hydrochloride. Journal of Pharmaceutical Sciences 92(4): 831-838.
[10] C.J. Strachan, T. Rades, D.A. Newnham, K.C. Gordon, M. Pepper, P.F. Taday, 2004. Using terahertz pulsed spectroscopy to study crystallinity of pharmaceutical materials. Chemical Physics Letters 390: 20-24.
[11] C. J. Strachan, P.F. Taday, D.A. Newnham, K.C. Gordon, J.A. Zeitler, M. Pepper, T. Rades, 2005. Using Terahertz Pulsed Spectroscopy to Quantify Pharmaceutical Polymorphism and Crystallinity. Journal of Pharmaceutical Sciences 94(4):837-846.
[12] Y. Ikeda, Y. Ishihara, T. Moriwaki, E. Kato, and K. Terada, 2010. A Novel Analytical Method for Pharmaceutical Polymorphs by Terahertz Spectroscopy and Optimization of Crystal Form at the Discovery Stage. Chem. Pharm. Bull. 58(1) 76-81.
[13] H-B. Liu, Y. Chen, X.-C. Zhang, 2007. Characterization of Anhydrous and Hydrated Pharmaceutical Materials with THz Time-Domain Spectroscopy. Journal of Pharmaceutical Sciences 96(4): 927-934.
[14] H.B. Liu, X.-C. Zhang, 2006. Dehydration kinetics of D-glucose monohydrate studied using THz time-domain spectroscopy. Chemical Physics Letters 429:229-233.
[15] M. D. King, W. D. Buchanan, T. M. Korter, 2011. Understanding the Terahertz Spectra of Crystalline Pharmaceuticals: Terahertz Spectroscopy and Solid-State Density Functional Theory Study of (S)-(+)0Ibuprofen and (RS)-Ibuprofen. Journal of Pharmaceutical Sciences 100(3):1116-1129.
[16] D.G. Allis, A.M. Fedor, T.M. Korter, J.E. Bjarnason, E.R. Brown, 2007. Assignment of the lowest-lying THz absorption signatures in biotin and lactose monohydrate by solid-state density functional theory. Chemical Physics Letters 440: 203-209)

[17] W. Limwikrant, K. Higashi, K. Yamamoto, K. Moribe, 2009. Characterization of ofloxacin-oxalic acid complex by PXRD, NMR, and THz spectroscopy. International Journal of Pharmaceutics 382: 50-55.
[18] M. Takahashi, Y. Kawazoe, Y. Ishikawa, H. Ito, 2006. Low-frequency vibrations of crystalline α, α-trehalose dehydrate. Chemical Physics Letters 429: 371-377.
[19] N. Nagai, R. Kumazawa, R. Fukasawa, 2005. Direct evidence of inter-molecular vibrations by THz spectroscopy. Chemical Physics Letters 413: 495-500.
[20] J. Shen, G. Wang, D. Jiang, L. Liang, X. Xu, 2010. Terahertz spectroscopic investigations of caffeine and 3-acetylmorphine. Optik 121 :1712-1716.
[21] Y. Watanabe, K. Kawase, T. Ikari, H. Ito, Y. Ishikawa, H. Minamide, 2004. Component anslysis of chemical mixtures using terahertz spectroscopic imaging. Optics Communication 234 : 125-129.
[22] R. Palermo, R. P. Cogdill, S. M. Short, J. K. Drennen III, P. F. Taday, 2008. Density mapping and chemical component calibration development of four-component compacts via terahertz pulsed imaging. Journal of Pharmaceutical and Biomedical Analysis 46 : 36-44.
[23] H. Wu, R.C. Lyon, A.S. Hussain, J.D. Drennen III, *et al.*. Application of Principal Component Analysis in Assessing Pharmaceutical Formulation Design: Exploring the Casual Links between the Tablet Processing Conditions and Drug Dissolution Rate. *Proceedings of Symposium on Innovations in Pharmaceutical and Biotechnology Development and Manufacturing, AIChE Annual Meeting* [TE001], pg. 635-642. AIChE Pub. No. 187, ISBN 0-8169-0929-6. San Francisco, Nov. 17-21, 2003.
[24] A.J. Fitzgerald, B.E. Cole, P.F. Taday, 2005. Nondestructive Analysis of Tablet Coating Thickness Using Terahertz Pulsed Imaging. Journal of Pharmaceutical Sciences 94(1):177-183.
[25] L. Ho, R. Müller, K.C. Gordon, P. Kleinebudde, M. Pepper, T. Rades, Y. Shen, P.F. Taday, J. A. Zeitler, 2008. Applications of terahertz pulsed imaging to sustained-release tablet film coating quality assessment and dissolution performance. Journal of Controlled Release, 127: 79-87.
[26] L. Ho, Y. Cuppok, S. Muschert, K.C. Gordon, M. Pepper, Y. Shen, F. Siepmann, J. Siepmann, P.F. Taday, T. Rades, 2009. Effects of film coating thickness and drug layer uniformity on in vitro drug release from sustained-release coated pellets: A case study using terahertz pulsed imaging. International Journal of Pharmaceutics, 382:151-159.
[27] L. Ho, R. Müller, C. Krüger, K.C. Gordon, P. Kleinebudde, M. Pepper, T. Rades, Y. Shen, P.F. Taday, J. A. Zeitler, 2010. Investigating Dissolution Performance Critical Areas on Coated Tablets: A Case Study Using Terahertz Pulsed Imaging. Journal of Pharmaceutical Sciences 99(1): 392-402
[28] J.A. Zeitler, L.F. Gladden, 2008. *In-vitro* tomography and non-destructive imaging at depth of pharmaceutical solid dosage forms. European Journal of Pharmaceutics and Biopharmaceutics, 71: 2-22
[29] C. Cahyadi, A.D. Karande, L.W. Chan, P.W.S. Heng, 2010. Comparative study of non-destructive methods to quantify thickness of tablet coatings. International Journal of Pharmaceutics, 398: 39-49
[30] T.J. Johnson, N.B. Valentine, S.W. Sharpe, 2005. Mid-infrared versus far-infrared (THz) relative intensities of room temperature *Bacillus* spores. Chemical Physics Letters 403:152-157.

[31] T.M. Korter, R. Balu, M.B. Campbell, M.C. Beard, S.K. Gregurick, E.J. Heilweil, 2006. Terahertz spectroscopy of solid serine and cysteine. Chemical Physics Letters 418: 65-70.
[32] N.N. Brandt, A.Yu. Chikishev, A.V. Kargovsky, M.M. Nazarov, O.D. Parashchuk, D.A. Sapozhnikov, I.N. Smirnova, A.P. Shkurinov, N.V. Sumbatyan, 2008. Terahertz time-domain and Raman spectroscopy of the sulfur-containing peptide dimmers: Low-frequency markers of disulfide bridges. Vibrational Spectroscopy 47: 53-58.
[33] N. Laman, S.Sree Harsha, D. Grischkowsky, and J. Melinger, 2008. High-Resolution Waveguide THz Spectroscopy of Biological Molecules. Biophysical Journal 94: 1010-1020.
[34] P.A. George, W. Hui, F. Rana, B.G. Hawkins, A.E. Smith, B.J. Kirby, 2008. Microfluidic devices for terahertz spectroscopy of biomolecules. Optics Express 16(3): 1577-1582.
[35] H.-B. Liu, G. Plopper, S. Earley, Y. Chen, B. Ferguson, X.-C. Zhang, 2007. Sensing minute changes in biological cell monolayers with THz differential time-domain spectroscopy. Biosensors and Bioelectronics 22:1075-1080.
[36]. E. Pickwell-MacPherson and V.P. Wallace, 2009. terahertz pulsed imaging-A potential medical imaging modality? Photodiagnosis and Photodynamics Therapy 6:128-134.
[37] H. Wu, E.J. Heilweil, A.S. Hussain, M.A. Khan, 2008. Process Analytical Technology (PAT): Quantification Approaches in Terahertz Spectroscopy for Pharmaceutical Applications. Journal of Pharmaceutical Sciences, 97(2): 970-984.
[38] P. Chalus, Y. Roggo, S. Walter, M. Ulmschneider, 2005. Near-infrared determination of active substance content in intact low-dosage tablets. Talanta 66:1294-1302.
[39] S.R.-Torres, J.D.P.-Ramos, K. R. Morris, E.R. Grant, 2005. Raman spectroscopic measurement of tablet-to-tablet coating variability. Journal of Pharmaceutical and Biomedical Analysis 38:270-274.
[40] C.F.Pereira, M.F. Pimentel, R.K.H. Galvao, F.A. Honorato, L. Stragevitch, M. N. Martins, 2008. A comparative study of calibration transfer methods for determination of gasoline quality parameters in three different near infrared spectrometers. Analytica Chimica Acta 611:41-47.
[41] Å. Rinnan, F. v. d. Berg, S.B. Engelsen, 2009. Review of the most common pre-processing techniques for near-infrared spectra. Trends in Analytical Chemistry 28 (10):1201-1222.
[42] Xie, L.,Wu, H., Shen, M., Augsburger, L., Lyon, R.C., Khan, M.A., Hussain, A.S.,and Hoag, S.W., 2008. Quality-by-Design (QbD): Effects of Testing Parameters and Formulation Variables on the Segregation Tendency of Pharmaceutical Powder Measured by the ASTM D 6940-04 Segregation Test. *Journal of Pharmaceutical Sciences* 97(10):4485-97.
[43] H. Wu and M. Khan, 2009. Quality-by-Design (QbD): An Integrated Approach for Evaluation of Powder Blending Process Kinetics and Determination of Powder Blending End-Point. *Journal of Pharmaceutical Sciences* 98(8):2784-2798.
[44] H. Wu, M. Tawakkul, M. White, and M. Khan. Quality-by-Design (QbD): An Integrated Multivariate Approach for the Component Quantification in Powder Blends. *International Journal of Pharmaceutics* 372(1-2):39-48, 2009.

[45] H. Wu and M. Khan, 2010. Quality-by-Design (QbD): An Integrated Process Analytical Technology (PAT) Approach for Real-Time Monitoring and Mapping the State of a Pharmaceutical Co-precipitation Process. *Journal of Pharmaceutical Sciences* 99(3): 1516-1534.
[46] H. Wu, M. White, M. Khan, 2011. Quality-by-Design (QbD): An Integrated Process Analytical Technology (PAT) Approach for a Dynamic Pharmaceutical Co-precipitation Process Characterization and Process Design Space Development. *International Journal of Pharmaceutics*, 405:63-78.
[47] H. Wu and M. Khan, 2011. Quality-by-Design (QbD): An Integrated Process Analytical Technology (PAT) Approach to Determine the Nucleation and Growth Mechanisms during a Dynamic Pharmaceutical Co-precipitation Process. *Journal of Pharmaceutical Sciences* 100(5): 1969-1986.
[48] ASTM Standard E1655-00. Standard Practices for Infrared Multivariate Quantitative Analysis.
[49] ASTM Standard D 3764-01: Standard Practice for Validation of Process Stream Analyzer Systems
[50] EMEA CPMP and CVMP Draft: Note for Guidance on the use of near infrared spectroscopy by the pharmaceutical industry and the data to be forwarded in part II of the dossier for a market authorisation. London, 14 Nov 2001, EMEA/CVMP/961/01; CPMP/QWP/3309/01 draft.
[51] R.D. Snee, 1977. Validation of regression models: Methods and examples, Technometrics 19:415-428.
[52] J. Michaelsen, 1987. Cross-validation in statistical climate forecast models, J. of Climate and Applied Meteorology 26:1589-1600.
[53] J. Obradovic, J. H.P. Collins, O. Hirsch, M. D. Mantle, M. L. Johns, L.F. Gladden, 2007. The use of THz time-domain reflection measurements to investigate solvent diffusion in polymers. Polymer 48: 3494-3503.
[54] Dykaar DR, Greene BI, Federici JF, Levi AFJ, Pfeiffer LN, Kopf RF., Appl Phys Lett 1991 59(3):262-4.
[55] H. Wu, E.J. Heilweil, A.S. Hussain, M.A. Khan, 2007. Process Analytical Technology (PAT): Effects of instrumental and compositional variables on terahertz spectral data quality to characterize pharmaceutical materials and tablets. International Journal of Pharmaceutics, 343: 148-158.
[56] B. Hancock, S. Shamblin, 1998. Water vapor sorption by pharmaceutical sugars. Pharmaceutical Science Technology Today, 1:345-351.
[57] R. Bird, W. Stewart, E. Lightfoot, 1960. Transport Phenomena. John Wiley, New York.
[58] A. Rahman, F. Stillinger, 1971. Molecular dynamics of liquid water. J. Chem. Phys. 55: 3336-3359.
[59] J. Hasted, S. Husain, F. Frescura, J. Birch, 1985. Far-infrared absorption in liquid water. Chem. Phys. Lett. 118: 622-625.
[60] L. Thrane, R. Jacobsen, P. Uhd Josen, S. Keiding, 1995. THz reflection spectroscopy of liquid water. Chem. Phys. Lett. 240:330-333.
[61] E. Castner, Jr., and Y. Chang, 1995. The intermolecular dynamics of liquid water. J. Chem. Phys. 102:653-659.

[62] H. Wu, A.S. Hussain, and M.A. Khan, 2007. Process Control Perspective for Process Analytical Technology: Integration of Chemical Engineering Practice into Semiconductor and Pharmaceutical Industries. *Chemical Engineering. Communications*. **194**: 760-779.
[63] US Food and Drug Administration. 2004. Innovation or Stagnation: Challenge and Opportunity on the Critical Path to New Medical Products.
http://www.fda.gov/oc/initiatives/criticalpath/whitepaper.html (accessed on 09/24/2012)
[64] J. Woodcock and R. Woosley, 2008. The FDA Critical Path Initiative and Its Influence on New Drug Development. *Annual Review of Medicine*, Vol. 59, DOI: 10.1146/annurev.med.59.09056.155819, Feb, 2008

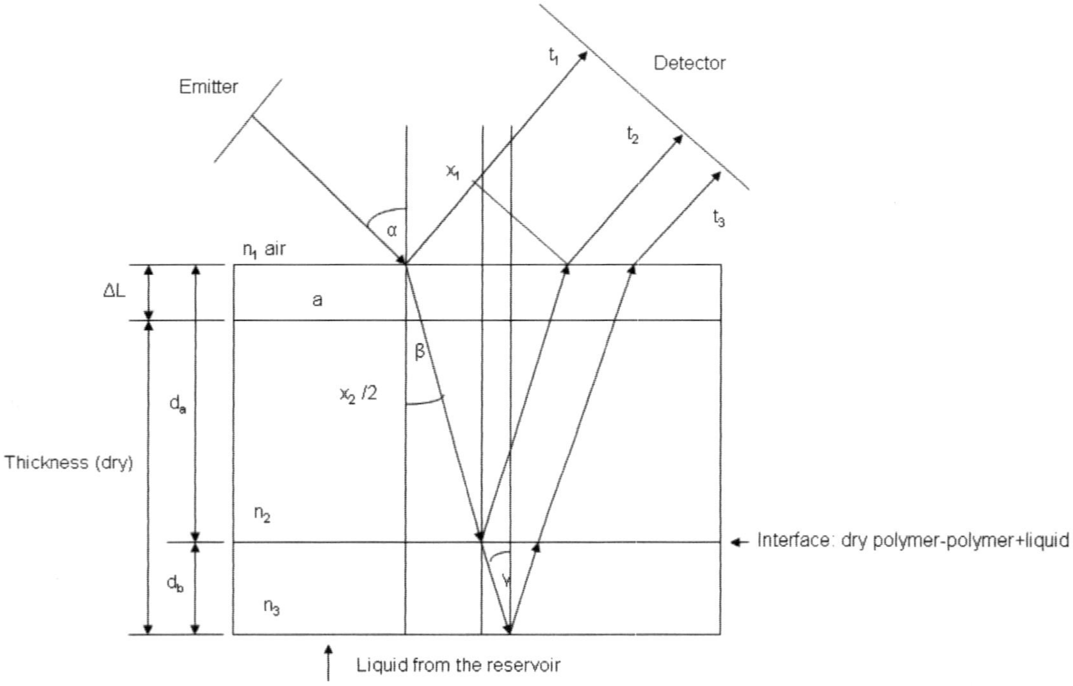

Figure 1. Figure 1 Schematic of the THz refection measurement from the polymer disc (adaption from Ref. [53] with kind permission from Elsevier)
n_1: index of refraction of air;
n_2: index of refraction of dry polymer;
n_3: index of refraction of the polymer penetrated by the liquid;
α: angle of incidence from air;
β: angle at which the THz beam is refracted when propagating from air into polymer disc;
a: shift of the reflection point;
x_1: path length difference in air;
x_2: path length difference in dry polymer;
d_a: thickness of the unpenetrated polymer layer;
d_b: thickness of the acetone-wet polymer layer;
ΔL: amount of swelling.

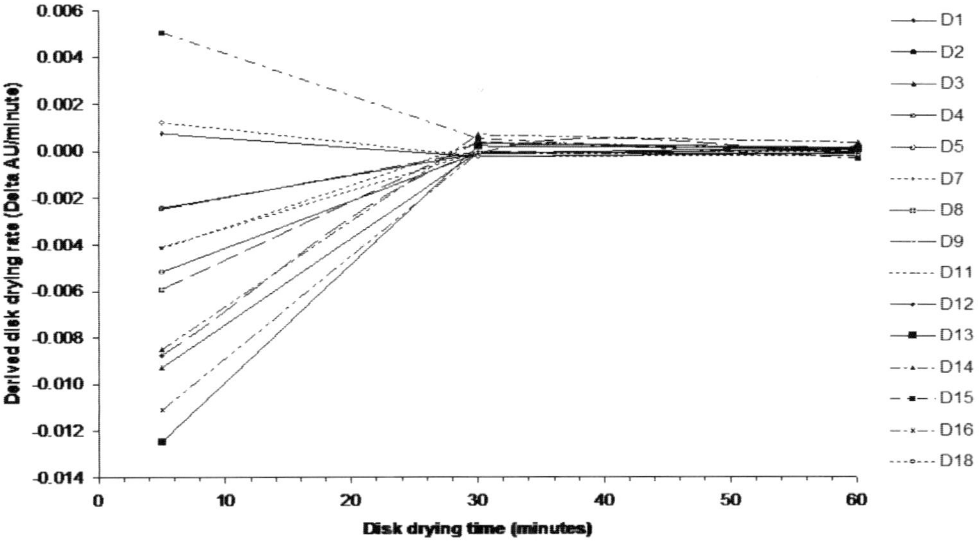

Figure 2(a)

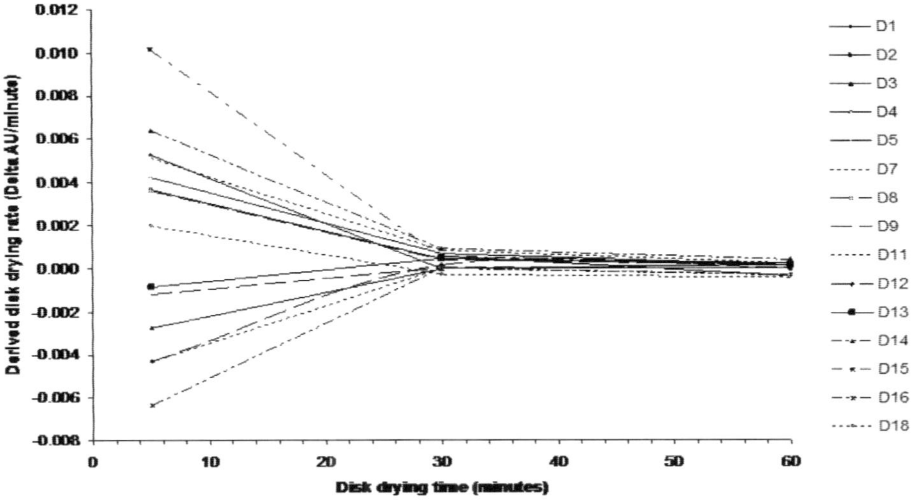

Figure 2(b)

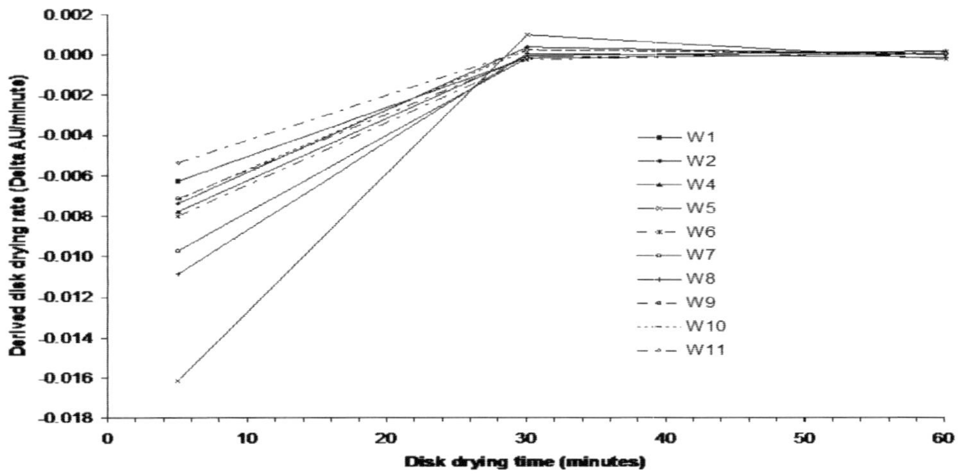

Figure 2(c)

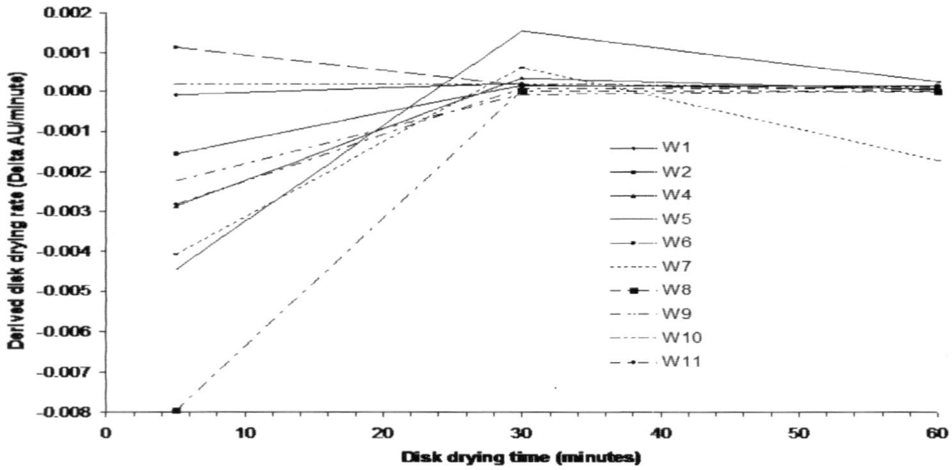

Figure 2(d)

Figure 2 Plots of derived disk drying rates vs. disk drying times based on absorbance data at THz bands around 54 cm^{-1} and 210 cm^{-1}. (a) Direct Compression (DC) tablets at 54 cm^{-1}; (b) DC tablets at 210 cm^{-1}; (c) Wet Granulation (WG) tablets at 54 cm^{-1}; (d) WG tablets at 210 cm^{-1}. In Figure 2(a)-(b), lengend D# stands for direct compression tablet #. In Figure 2(c)-(d), lengend W# stands for wet granulation tablet #. (adaption from Ref. [55] with kind permission from Elsevier)

(146a) A Model-Based Approach for the Construction of Design Spaces in Quality-by-Design

Masako Kishida, University of Illinois at Urbana-Champaign, Urbana, IL and **Richard D. Braatz**, Chemical Engineering, Massachusetts Institute of Technology, Cambridge, MA

This presentation considers the characterization of the set of allowable real parameter uncertainties for the output of a nonlinear system to be located in a prespecified target range. This word "parameter" is used by FDA and in this presentation to refer to any variable held at a fixed value during a single experiment. For example, this could be a parameter in a model, an initial condition such as the initial concentrations in a semibatch chemical reactor, or an operational parameter such as the cooling rate in batch reactor.

This uncertainty analysis problem is motivated by various industrial initiatives for Quality by Design (QbD) (Juran 1992), which have the objective of designing a manufacturing process so as to ensure that the product achieves its desired quality qualifications despite perturbations in the manufacturing process. The QbD literature refers to the set of allowable parameters as the design space (Juran 1992), which in industrial practice is constructed from the data collected from a large number of time-consuming and expensive experiments (e.g., see Garcia-Munoz et al, 2010, for a description of the state-of-the-art in industrial practice and Togkalidou et al, 2001, for an example of experimental design for generating models for pharmaceutical processes). This presentation considers the alternative construction of the design space based on a first-principles, grey-box, or black-box model of the manufacturing process, as fewer experiments are required to construct a model than to directly construct a design space from experiments.

In the control literature, this design space is typically characterized in terms of a robustness margin (Safonov and Athans 1977). Numerous publications quantify the effects of a specified norm-bounded perturbation in a system model on a signal norm on the system output (Wang et al 1992; Zhou et al, 1995) or on outer bounds for state or output trajectories (Durieu et al 2001; Kishida and Braatz 2010, 2011; Kishida et al 2011; Scott and Barton, 2009). The structured singular value and quadratic stability margin quantify the amount of uncertainty allowed to maintain closed-loop stability or a signal norm characterizing output performance (Ma et al 2001, 2002; Hovd et al 2003; Zhou et al 1995; Boyd et al 1997), for linear time-invariant or time-varying systems or linear systems coupled with bounded nonlinear perturbations. This presentation considers nonlinear systems and defines allowable variations in parameters for instantaneous specified bounds on the states or outputs to be satisfied, as this characterization of output performance is used in QbD applications (Juran 1992). The overall approach applies to both open- and closed-loop systems. As the exact generation of a design space is a challenging computational problem for a large number of parameters (Braatz et al 1994), tight upper bounds are determined.

The system output is approximated by a polynomial function (Taylor series expansion) or a rational function (Pad´e approximant) followed by application of the skewed structured singular value to obtain a set of allowable real parameters that ensures that the system output will be in a target set. A target set described by upper and lower bounds on each output is considered here; the same analysis can be used for an ellipsoidal target set by slight modification of the skewed structured singular value analysis (Smith 1990). Both polynomial and rational approximants can be written in the form of a linear fractional transformation, which is used in the next step of the analysis of applying the skewed structured singular value to quantify the allowable uncertainty set. Algorithms are also discussed that generate the algebraic constructions needed by the method as well as model reductions to reduce computational cost (Russell et al 1997; Russell and Braatz 1998).

Several design spaces are constructed for models reported by pharmaceutical companies, including a crystallization operation and a nasal spray.

References:

S. Boyd, L. El Ghaoul, E. Feron, and V. Balakrishnan. Linear Matrix Inequalities in System and Control Theory.

SIAM Press, Philadelphia, PA, 1997.
R. D. Braatz, P. M. Young, J. C. Doyle, and M. Morari. IEEE T. Automat. Contr., 39:1000–1002, 1994.
C. Durieu, E.Walter, and B. Polyak. J. Optimiz. Theory App., 111:273–303, 2001.
S. Garcia-Munoz, S. Dolph, and H. W. Ward. Comput. Chem. Eng., 34:1098–1107, 2010.
M. Hovd, D. L. Ma, and R. D. Braatz. Ind. Eng. Chem. Res., 42:2183–2188, 2003.
J. M. Juran. Juran on Quality by Design: The New Steps for Planning Quality into Goods and Services. The Free Press, New York, 1992.
M. Kishida and R. D. Braatz. Opt. Contr. Appl. Met., 31:433–449, 2010.
M. Kishida and R. D. Braatz. In Proc. of IEEE CDC/ECC, pages 5671–5676, Orlando, FL, 2011.
M. Kishida, P. Rumschinski, R. Findeisen, and R. D. Braatz. Efficient polynomial-time outer bounds on state trajectories for uncertain polynomial systems using skewed structured singular values. Proc. of the Joint Symposium on Computer-Aided Control System Design and Systems with Uncertainty, Denver, Colorado, 216-221, 2011
D. L. Ma and R. D. Braatz. IEEE Trans. Control Syst. Tech., 9:766–774, 2001.
D. L. Ma, J. G. VanAntwerp, M. Hovd, and R.D. Braatz. IEE Proc. - Control Theory & Applications, 149:423–432, 2002.
E. L. Russell and R. D. Braatz. Comput. Chem. Eng., 22:913–926, 1998.
E. L. Russell, C. P. H. Power, and R. D. Braatz. Int. J. Robust Nonlin. Control, 7:113–125, 1997.
M. G. Safonov and M. Athans. IEEE TAC, 22:173–179, 1977.
J. K. Scott and P. I. Barton. Comput. Chem. Eng., 34:717–731, 2009.
R. S. R. Smith. Model Validation for Uncertain Systems. PhD thesis, California Institute of Technology, Pasadena, CA, 1990.
T. Togkalidou, R. D. Braatz, B. K. Johnson, O. Davidson, and A. Andrews. AIChE J., 47:160–168, 2001.
Y. Wang, L. Xie, and C. E. de Souza. Syst. Control Lett., 19:139–149, 1992.
K. Zhou, J. C. Doyle, and K. Glover. Robust and Optimal Control. Prentice Hall, 1995.

(146b) Fluid Bed in a Flash

Otto Scheibelhofer[1], Roland Hohl[1], Sharareh Salar-Behzadi[1], Detlev Haack[2], Kai Koch[3], Stephan Sacher[1], Jose C. Menezes[4] and Johannes G. Khinast[5,6], (1)Research Center Pharmaceutical Engineering, Graz, Austria, (2)Hermes Arzneimittel GmbH, Munich, Germany, (3)Innojet Herbert Hüttlin, Steinen, Germany, (4)Research Center for Pharmaceutical Engineering, Graz, Austria, (5)Institute for Process and Particle Engineering, Graz University of Technology, Graz, Austria, (6)Research Center Pharmaceutical Engineering GmbH, Graz, Austria

Fluid Bed in a Flash

O. Scheibelhofer[1], Roland Hohl[1], Sharareh Salar Behzadi[1], Detlev Haack[2], Kai Koch[3], Stefan Sacher[1], Jose C. Menezes[1], J. G. Khinast[1,4]

[1] Research Center Pharmaceutical Engineering GmbH, Graz, Austria

[2] Hermes Arzneimittel GmbH, Munich, Germany

[3] Innojet Herbert Hüttlin, Steinen, Germany

[4] Institute for Process and Particle Engineering, Graz University of Technology, Graz, Austria

Email for correspondence: otto.scheibelhofer@rcpe.at (O. Scheibelhofer), khinast@tugraz.at (J.G. Khinast)

Fluid Beds are a versatile and widely used unit operation in the pharmaceutical industry. Putting to use their property of very large particle to air contact surface, they are an advanced step for drying, coating and more. A great deal of interest has already been invested, concerning the properties of fluid beds, especially their geometry and flow pattern. Mostly, this is a field for simulations.

Nonetheless, the efficiency of fluid beds is often driven by the experience of their operators. This is a result of fluid beds being very sensitive to subtle changes, which can affect the flow pattern significantly, even dis-enabling the fluid bed process as a whole. Even more so, this increases the difficulty of placing sensors in such environment, making systematic experimental investigations an exhaustive work.

To overcome some of these difficulties, near-infrared spectroscopy seems a natural choice, for two reasons. First, the method is non-invasive, or even contactless, so to avoid the disturbance of the established flow pattern. Secondly, the method can be very fast and can therefore cope with the time-scale of flow inside the bed.

A single NIR probe is used for monitoring the fluid bed and following the trajectory of our process, for this a process NIR from Sentronic is used.

However, as there are different zones of flow and progress inside the bed, a single probe is not enough, to get an idea of the overall maturity inside the fluid bed.

Here we used a new spectrometer prototype provided by EVK, Raaba, Austria in order the monitor a coating process. The prototype is based on a push-broom principle, but instead of a line, a standardized port plate enables the mapping of different near-infrared probes, via a prism-grating-prism system onto the detection chip. As a consequence, several positions inside the fluid bed can be monitored simultaneously, with a frequency of several Hz. even more, the records of the probes can be directly compared, as they are guaranteed to be recorded at the same time, and there is no spectrometer drift in between probes, etc.

The probes are distributed inside the fluid bed vessel at carefully chosen positions, at the wall, in order to not disturb the flow pattern, but give at the same time the most relevant information.

The information gained is of value twofold. First, it tells about the process itself, and can be used to estimate necessary process time and quality. Second, it tells about the mechanisms inside the fluid bed, spatial differences, etc.; which is very valuable for final probe positioning, the task of up-scaling, and more.

Estimation of the sample positions and the process is done with risk analysis, and design of experiments is also used for generating an experimental setup pattern. Within the multiprobe measurement system, we will also evaluate the sample position of the single probe.

Figure 1: The Ventilus 2.5 fluid bed reactor is shown., with the Sentronic NIR probe attached.

(146c) From Experimental Data to Building Integrated Dynamic Flowsheet Models for Pharmaceutical Processes

Fani Boukouvala[1], Fernando J. Muzzio[2] and Marianthi Ierapetritou[2], (1)Chemical and Biochemical Engineering, Rutgers University, Piscataway, NJ, (2)Department of Chemical and Biochemical Engineering, Rutgers, The State University of New Jersey, Piscataway, NJ

FDA harmonization guidelines [1] aim to promote building quality into design of pharmaceutical processes through mechanistic understanding of powder process behavior. Based on this paradigm, the industry is moving with a fast pace towards a model-based process and product design, which will undoubtedly lead to competitive manufacturing and better product quality [2]. However, process modeling is highly dependent on experimental or computational data, either in the form of empirical correlations or during the stage of parameter estimation. At the same time, there has been a high interest in the population of a flowsheet model library for solids-based pharmaceutical processes. Flowsheet modeling has been an imperative process engineering tool for the design and optimization of liquid-based processes [3-4]. This work aims to summarize all of the identified challenges that originate from the incorporation of experimental and computational data into a dynamic flowsheet model library for pharmaceutical processes, in order to develop a generic model library which may be used for a wide range of raw materials and processes, integrated to form complete production lines.

Many challenges arise due to the nature of pharmaceutical materials and processes compared to other industries, such as increased variability of powder material properties and operating conditions, uncharacterized materials, lack of well-established correlations of material properties and process operation and oftentimes lack of predictive models for certain processing steps. However, during the recent years where there has been a strong focus in enhancing process understanding, large amounts of data for the characterization of processes or materials have been collected. In certain cases, these data have provided insight about the governing underlying mechanisms and have lead to the development of first-principle models, while in other cases data have been used to build empirical response surface models, or multivariate latent variable models in cases of high dimensionality. In fact, even in the well-established liquid-based flowsheet models, certain physical phenomena which are insufficiently understood, such as chemical reaction kinetics or heat and mass transfer effects are represented through empirical correlations which are incorporated into first-principle models to create hybrid or semi-empirical modules [5].

Moreover, the complex nature of powder based-processes has also lead to the development and use of high-fidelity spatially distributed simulation models that aim to capture the complex behavior of interacting particles in order to provide insight into detailed critical material attributes such as particle size distributions. This introduces another challenge towards the final goal of flowsheet model building, which is the incorporation of multiscale computational data, derived from very expensive computer models. Fortunately, however, through this work it will be shown that the handling of such computational data is very similar to that of experimental data.

Finally it should be mentioned that the link between data and models is always a closed-loop process, since even after the development of the necessary models, the need of experimental data is once again imperative for the estimation of the model parameters. In this case the experimental design will be guided by the models, in order to ensure that they contain the necessary information needed for the parameter estimation.

A great opportunity arises in order to efficiently incorporate the information obtained from large data-bases and build and/or refine the flowsheet model library following well-established techniques of the process systems engineering community. This work aims to discuss the state of the art in first-principle models used to simulate pharmaceutical manufacturing and their proposed combinations and modifications in order to maximize the captured information through incorporation of useful data-based correlations. Most importantly, the need to consider each process as part of a large integrated puzzle should be considered even from the early stages of experimental design, since due to the need for simulation of continuous integrated lines, the propagation of a common set of material properties of powder mixtures is essential. The challenges and proposed remedies will be shown through specific case studies for a variety of manufacturing scenarios of pharmaceutical tablets such as direct compaction, dry and wet granulation.

References:

1. *Q8 Pharmaceutical Development*. 2006; Available from: http://www.fda.gov/downloads/RegulatoryInformation/Guidances/ucm128029.pdf.

2. Oksanen, C.A. and S. García Muñoz, *Process modeling and control in drug development and manufacturing.* Computers and Chemical Engineering, 2010. 34(7): p. 1007-1008.

3. Boukouvala, F., et al., *An integrated approach for dynamic flowsheet modeling and sensitivity analysis of a continuous tablet manufacturing process.* Computers and Chemical Engineering, In Press, 2012.

4. Werther, J., C. Reimers, and G. Gruhn, *Design specifications in the flowsheet simulation of complex solids processes.* Powder Technology, 2009. 191(3): p. 260-271.

5. Kahrs, O. and W. Marquardt, *The validity domain of hybrid models and its application in process optimization.* Chemical Engineering and Processing: Process Intensification, 2007. 46(11): p. 1054-1066.

(146d) Modeling and Simulation of a Wurster Column for Particulate Coating Using Population Balance Moldels and Thermodynamics

Jianfeng Li, Chemical Engineering, Purdue University, West Lafayette, IN, **Salvador García-Muñoz**, Process Modeling & Engineering Technology, Pfizer Worldwide Research & Development, Groton, CT, William R. Ketterhagen, Pfizer Global Research and Development, Pfizer Inc., Groton, CT and James D. Litster, School of Chemical Engineering, Department of Industrial and Physical Pharmacy, Purdue University, West Lafayette, IN

The Wurster column coating process of pharmaceutical solids has been studied for many years and a considerable body of literature has been produced covering various aspects of the process. To achieve uniform active coating content is one of the key interests for the pharmaceutical manufacturing industry. Thus, a process model for predicting the evolution of a granule size distribution from first principles is desired.

In this work, a compartment based population balance model, combined with thermodynamic coating model has been developed to model a pharmaceutical coating process in the Wurster column coater. The model can predict certain key process parameters and the dynamic particle size distribution, and it can also be used to simulate responses of the system to a change in the working conditions.

In many industrial practices, the coating equipment has been identified into coating/non-spray regions. It is believed that both the variation of residence time in the spray region and the variation of cycle time lead to the non-uniform coating mass distribution among individual particles. Crites and Turton (2005) have developed a compartment based model to incorporate the cycle time distribution feature for the coating process in the Wurster column coater. The model was composed of fundamental flow types such as plug flow and well-mixed flow regions to describe phenomena including dead zone and side pass. Other instances include a series of well-mixed tanks model structure for a top-spray fluidized bed coater (Ronsse 2005), a twin region compartmental model for a rotating drum coater (Dennis et al. 2001).

In this work we use a population balance (PB) approach to model the particle size distribution of the coated particulates as a function of time. While the particle growth due to coating strongly relates to the number and duration particles passing through the spray region, a practical way to constructing PB models is based on the decomposition of the coater into spray/non-spray compartments. Thermodynamics of the spray are evaluated with a simple energy balance (see am Ende et al. 2005)

In terms of the flow of particles across the unite, the Wurster column coater is divided into three compartments including an ideal plug flow region, an ideal fountain CSTR, and a non-ideal annulus CSTR (in a similar way to that proposed by Crites and Turton (2005)).

1. The plug flow compartment refers to the region enclosed by the draft-tube, where particles are highly fluidized and heated by the upflow hot air. The coating solution is assumed to be delivered on to these particles from the spray nozzle centered in the middle of the distribute plate.

2. The ideal CSTR compartment represents the fountain region, in which particles enter into the chamber and continuously receive spray.

3. Particles are then settled down towards the bottom of the coater in the annulus region, described by the non-ideal CSTR. Particles spend most of the time in this region. The non-ideality refers to the factor that smaller particles have longer recycle time compared to larger particles (Liu and Litster 1993), so the characteristic time of the CSTR for the annulus region is a function of particle size.

Experimental data is used to determine the thermal properties of the column as well as growth kinetics. Predicted values of particle size as a function of time were found in good agreement with particle size distribution measurements (by laser diffraction) of samples taken throughout the process.

References

am Ende, M.T. and Berchielli, A., 2005. A thermodynamic model for organic and aqueous tablet film coating. Pharmaceutical Development and Technology 1, 47–58.

Crites, T and Turton, R., 2005. Mathematical model for the prediction of cycle-time distributions for the Wurster column-coating process. Ind. Eng. Chem. Res. 44, 5397-5402.

Denis, C., Hemati, A., Chulia, D., Lanne, J.Y., Buisson, B., Daste, G., Elbaz, F., 2003. A model of surface renewal with application to the coating of pharmaceutical tablets in rotary drums. Powder Technology 130,174-180.

Liu, L. X., Litster, J. D., 1993. Coating mass-distribution from a spouted bed seed coater - experimental and modeling studies. Powder Technology 74, 259-270.

Ronsse, F., Pieters, J.G., Dewettinck, K., 2007. Combined population balance and thermodynamic modelling of the batch top-spray fluidised bed coating process. Part I—Model development and validation. Journal of Food Engineering 78, 296–307.

(146e) Exceptional Event Managemnt Applied to Continuous Pharmaceutical Manufacturing

Anshu Gupta, Gintaras V. Reklaitis and Venkat Venkatasubramanian, Chemical Engineering, Purdue University, West Lafayette, IN

One of the important challenges in effective real time process management is the implementation of intelligent systems that can assist human operators in making supervisory control decisions, instead of simply sounding an alarm when process variables go out of range. Operator failures to exercise the appropriate mitigation actions often have an adverse effect on product quality, process safety, occupational health and environmental impact. The economic effect of such exceptional events is immense; an estimated $20billion/year losses in petrochemical industry have been reported. The challenges and opportunities for improvements are even larger in the pharmaceutical manufacturing domain because so much of the processing involves particulate and granular systems whose characteristics tends to be more problematic than that of fluids. The development and manufacturing of pharmaceutical products are governed by strict safety regulations but with the advent of Process Analytical Technology (PAT) initiative advanced by the FDA; gives the pharmaceutical industry an opportunity to apply various systems engineering tools. Early detection and diagnosis of process faults while the plant is still operating in a controllable region can help avoid abnormal event progression, production disruptions and productivity losses. An EEM framework has been developed to deal with fault detection, diagnosis and mitigation of conditions that result from process anomalies. The framework developed uses a combination of Principal Component Analysis (PCA), Signed Digraphs (SDG) and Qualitative Trend Analysis (QTA) in applications involving continuous pharmaceutical product manufacturing line. An ontological database has been created to maintain records of the signatures of a number of exceptional events typical of a dry granulation line and their associated mitigation strategies. In trials on pilot scale equipment we have found that the EEM framework was able to detect and diagnose several types of faults within a few seconds of their inception and to provide mitigation advisories to the operator.

(146f) Modelling of a Filter Drying Process

Giovanni Giorgio[1], Jason Mustakis[2], **Mark A. Pinto**[3] and Sean K. Bermingham[3], (1)Particle Generation and Control Engineering, GlaxoSmithKline Ltd., Stevenage, United Kingdom, (2)Pfizer Global Research & Development, Pfizer Inc., Groton, CT, (3)Process Systems Enterprise, London, United Kingdom

Filter drying, also referred to as vacuum contact drying, is a process widely used in the pharmaceutical industry to remove water or organic solvents. The advantage of this process over other drying operations is that it can be used to dry solids that are sensitive to oxygen and temperature.

The entire drying process is typically carried out in two stages in a vessel. Typically a cylindrical vessel is used although conical vessels are also encountered. In the first blowdown stage, pressure in the form of nitrogen is applied to remove excess solvent by convective drying. At the end of this stage, the wet filter cake typically has around 45% solvent on a wet basis.

The second, longer, period involves heating the walls of the filter cake while vacuum is applied to dry the cake to about 1% to 0.1%. During this stage, heat conduction from the wall to the interior of the cake is often the rate limiting step. While the process can be carried out without agitating the bed, this can lead to excessively long drying times and heterogeneity of the cake. On the other hand, continuous long agitation often results in agglomeration and/or breakage of the solid. As a compromise, intermittent periods of agitation are carried out to redistribute solvent and heat, resulting in faster drying times but without significant agglomeration and/or breakage of solid taking place.

A detailed model of the filter dryer unit operation has previously been published in the literature (Murru et al, 2011). The key features of the model are:

1. The second drying phase is modelled and drying is assumed to take place primarily due to heat conduction from the equipment walls. The blowdown phase where convective drying takes place is not accounted for.

2. A single solvent is considered. This allowed for process optimisation by allowing different solvents to be examined for the effects on total drying times.

The model was successfully used to predict drying times for various equipment sizes and solvents at both pilot plant and manufacturing scales.

In this work, the model is further extended to consider the following phemonena.

1. Convective drying is considered. This allows the modelling of the initial blowdown drying phase. The effects of the applied suction during the second drying phase are also explicitly accounted for. The application of hot nitrogen flow over the bed can significantly reduce the time of drying.

2. The model is extended to non-ideal mixtures of solvents. The presence of solvent mixtures in the drying operation has a significant effect on the drying time but most importantly lead to change of composition of the residual solvent during drying. This can lead to stability issues (hydrates / solvates) or can lead to particle bridging and agglomeration. Understanding the dynamics of the solvent composition is crucial to avoid these situations. The ability to handle multiple solvents simultaneously will allow for greater flexibility in selecting solvents for process optimisation.

3. Modelling of the vapour processing step is undertaken. This allows for data from downstream condensation and scrubbing unit operations to be used in model validation.

References

M Murru, G Giorgio, S Montomoli, F Ricard, F Stepanek (2011) Model-based scale-up of vacuum contact drying of pharmaceutical compounds. *Chemical Engineering Science* **66**, 5045-5054

(182b) Case Studies Involving Mechanistic and Empirical Models Used to Define Design Space Ranges and Corresponding Large Scale Verification

Jose E. Tabora, Chemical Development, Bristol Myers Squibb, New Brunswick, NJ and Daniel Hallow, Bristol-Myers Squibb Company, New Brunswick, NJ

Placeholder only follows. Abstract to be added.

Abstract Text guidelines: A minimum of 150 words is required for the abstract text, and no more than 1500 words may be submitted. Each abstract is allowed to include up to 20 figures.
From http://www.pharmaqbd.com/bms_tabora_aiche_award_2011/
There is some resistance to the adoption of MVA modeling stemming from the prolific reliance on one-factor-at-a-time experimentation. While this approach is often appropriate and acceptable in early stages of process development, it frequently continues to be utilized in later stages of development although increased efficiencies may be attained through the implementation of combined MVA modeling/experimental approaches to improve process understanding and overcome key development challenges.
In addition, the discussion of which parameter(s) to explore simultaneously can be nuanced and complex, thereby compounding the reluctance to incorporate multivariate experimentation into process development and optimization workflows. However, I am very optimistic about the potential for increased acceptance and uptake of such approaches and have seen the adoption of these techniques increase substantially over the past five years. It is not only a regulatory expectation, but also leads to substantial gains in efficiency, which our scientists are recognizing quite rapidly.
It is widely acknowledged in the business intelligence sector that efficient multivariate analysis and modeling generally require significant efforts in data aggregation. The same holds true in chemical process development, and although tools to support this requirement are not currently readily available, there is an increasing appreciation of the importance of data consolidation. Hence, the industry is aggressively moving toward solutions to address this gap through the use of local databases and datamarts that aggregate data from multiple sources.

(182c) The Role of PAT, Condition Monitoring and Adaptive Control in Delivering QbD

Steve Hammond, Pfizer, Inc., Morris Plains, NJ and Mark Smith, Pfizer, Inc., Peapack, NJ

"The Role of PAT, Condition Monitoring and Adaptive Control in Delivering QbD"

Stephen Hammond, Mark Smith, Pfizer Global Supply, Peapack NJ, USA

The use of spectroscopic techniques such as NIR spectroscopy to deliver greater process understanding, either providing difficult to obtain compositional data or more data across a process has been a key factor in the expansion of QbD. The potential for condition monitoring of a unit operation in real-time during the manufacturing process adds enormously to the ability to understand, and reduce the risk of any unit operation.

This talk will describe the spectroscopic techniques used to assess the process during qualification, and the measurements applied to develop a RtR filing for a low dose product. The RTR approach has been approved in Europe and by FDA. Also in many other countries around the world.

The future state of QbD will be proposed, where modeling and control based on optimum process trajectories can be effective in reducing variation due to ingredients or processing equipment, that effects the economics of producing product.

(189a) Assessing Excipient Variability within QbD Drug Product Development

Joseph Kushner IV, Drug Product Design, Pfizer Worldwide Research and Development, Groton, CT

Understanding variability in excipient physico-chemical properties is becoming an important aspect of Quality-by-Design (QbD) drug product development. However, present experimental methods are only able to study a few physico-chemical properties on a few excipient lots due to time, cost, and sample gathering considerations. To improve the robustness of drug products to excipient variability, analysis methods are needed to examine a broader range of excipient physico-chemical properties. An alternative analysis method is proposed which uses quantitative physico-chemical property data gathered by excipient vendors to evaluate excipient lot-to-lot variability in a comprehensive, low cost manner. Microcrystalline cellulose, spray-dried lactose, and magnesium stearate were selected as commonly-used excipients for this demonstration. The proposed analysis method, which is based on multivariate approaches, offers drug product developers several advantages over present experimental methods, including the ability to: 1) examine excipient products for manufacturing site and/or year-to-year variations, 2) quantify a domain of prior experience for each excipient by determining the percentage of excipient lots contained within a multi-dimensional ellipsoid described by the excipient lots used during drug product development, and 3) rationally select excipient lots from the vendors inventory to maximize the domain of prior experience throughout the drug development process. Case studies will be reviewed illustrating successful application of this method supporting drug product robustness to critical excipient material properties. Leveraging prior knowledge from excipient vendors' experience with excipient lot-to-lot variability should be an integral part of a QbD drug product development program to understand the role of excipient variability on the robustness of drug product manufacturability and performance.

(189b) Inversion of Latent Variable Regression Models to Support Product and Process Development

Emanuele Tomba[1], Massimiliano Barolo[1] and Salvador García-Muñoz[2], (1)DII - Department of Industrial Engineering, University of Padova, Padova, Italy, (2)Process Modeling & Engineering Technology, Pfizer Worldwide Research & Development, Groton, CT

When designing a new product and its manufacturing process, there is the need to understand how the raw material properties and the process parameters interact and affect the product performances, in order to control them and optimize the manufacture. For these reasons, usually large amounts of experiments are carried out to independently excite all the driving forces acting on the system, in order to gain a deep understanding of the relations between inputs (raw materials, process parameters) and outputs (final product properties). Even if systematically designed, the number of experiments to be performed may become prohibitive, especially if a large number of candidate materials and permutations have to be considered, as for the formulation of a pharmaceutical product. If data on historically developed products were available, useful information could be drawn from these databases, to support the design and the optimization of new products and of their manufacturing processes, with the aim of accelerating the development cycle guiding the experimentation from the first steps.

Latent variable regression models (LVRMs) are tools specifically designed to analyze large datasets of highly correlated data reducing the vast information in them in a small number of variables (the *latent variables*) which indentify the underlying driving forces relating input data with system outputs. The driving forces identified from the available historical databases represented by the LVRM equations can then be used to support the design of new products: since the design activity aims at finding the best combination of inputs (raw material properties and process conditions) which ensure a desired product profile in output, the design objectives could be achieved through a LVRM inversion operation.

In this contribution we show how LVRM inversion [1,2] can be feasibly used to support pharmaceutical product and process development activities. Since the LVRM inversion can have several solutions depending on the rank of the historical datasets used to build the model, the inversion has to be posed as an optimization problem [3,4]. A framework is therefore proposed to manage the different constraints and objectives the design problem may have, in order to solve in each situation the most appropriate LVRM inversion problem.

The proposed procedure is tested on two different product design applications, related to a particle engineering problem for the design of a wet granulated product of desired properties and to the selection of the best formulation for a given API, in order to obtain blends suitable for direct compression. In both cases the LVRM inversion is shown to be very effective in estimating the best inputs to obtain the desired product in output, giving a systematic tool to guide experiments since the first stages of development for the establishment of the system design space.

References

[1] Jaeckle, C. M. and J. F. MacGregor (1998). Product design through multivariate statistical analysis of process data. *AIChE J.*, **44**, 1105-1118.

[2] Jaeckle, C. M. and J. F. MacGregor (2000). Industrial applications of product design through the inversion of latent variable models. *Chemom. Intell. Lab. Sys.*, **50**, 199-210.

[3] García-Muñoz, S., T. Kourti, J. F. MacGregor, F. Apruzzese and M. Champagne (2006). Optimization of batch operating policies. Part I. Handling multiple solutions. *Ind. Eng. Chem. Res.*, **45**, 7856-7866.

[4] García-Muñoz, S., J. F. MacGregor, D. Neogi, B. E. Letshaw and S. Mehta (2008). Optimization of batch operating policies. Part II. Incorporating process constraints and industrial applications. *Ind. Eng. Chem. Res.*, **47**, 4202-4208.

(189c) Enabling Predictive Catalyst and Reaction Design in Asymmetric Organocatalysis (Invited)

Donna G. Blackmond, Dept. of Chemistry, The Scripps Research Institute, La Jolla, CA

Over the past several decades, the burgeoning field of asymmetric organocatalysis has introduced novel solutions to a number of transformations that construct carbon-carbon and carbon-heteroatom bonds in organic synthesis directed toward medicinal chemistry and drug development. This work has produced a wealth of experimental observations – primarily concerning product selectivity – including the exciting concept of cascade reactions where consecutive reactions fix multiple stereocenters in one pot. The area is now poised to emerge from empirical discovery to prediction and design of reaction outcomes enabling practical application. This next stage requires an increased focus on the underlying reaction mechanisms that afford efficient and selective transformations. Understanding complexity in competitive reactivity patterns in these systems is key to the development of design tools for improving both reactivity and selectivity in known reactions as well as for enabling transformations that have eluded us to date.

Our goal is to enable predictive catalyst and reaction design by developing experimental kinetic and spectroscopic protocols to probe complex reaction pathways. Investigations to date have relied primarily on empirical optimization based on simple analysis of final reaction product outcomes, which may neglect the importance of reaction dynamics and the interplay between stability and reactivity of intermediates in competitive networks. The temporal monitoring of reaction and selectivity profiles combined with spectroscopic identification of dynamic intermediate species is an underutilized approach in organocatalysis. Its development will aid in the delineation, rationalization, and exploitation of unexpected modes of stereocontrol. In this work we explore the generality of a proposed Curtin-Hammett paradigm for different classes of organocatalysts, applying comprehensive kinetic and spectroscopic protocols to a series of key transformations in pyrrolidine-based organocatalytic systems involving enamine intermediates. The central theme is the incorporation of our pioneering approach to quantitative kinetic analysis into a range of mechanistic tools. Kinetic-assisted mechanistic analysis will be developed as an orthogonal tool for discovery and optimization by rational design based careful documentation of reactivity patterns. This approach will address unexplained features of existing reactions as well as enable design of new catalyst/substrate combinations for high efficiency and selectivity, including complex reaction networks with multiple sequential transformations that begin to resemble metabolic systems, whence the inspiration for organocatalysis originates.

(189d) Implementing a Recipe Based Framework to Support Product Lifecycle Management

Adam Fermier[1], John Cunningham[2], Thomas Difeo[3], John Dingerdissen[4], James Kenyon[3], Steve Mehrman[3], Terry Murphy[5], Eugene Schaefer[2], Tom Schultz[6], Brian Sherry[7], James Weber[3] and Paul McKenzie[7], (1)Janssen Pharmaceuticals, Spring House, PA, (2)PDMS, Malvern, (3)PDMS, Spring House, (4)PDMS, Radnor, (5)PDMS, Cork, Ireland, (6)Janssen, Titusville, (7)Janssen Supply Chain, Horsham

The pharmaceutical industry to date has not fully leveraged data analysis methods such as multivariate techniques, e.g., multivariate statistical process controls, as achieved in some other industries. It has been postulated that the lack of such analysis was in caused in part, by the ability of the pharmaceutical industry to absorb sub-optimal / inefficient processes. We contend it is due to a more systematic challenge in the industry related to streamlined and effective data management.

The data management challenge stems from the diversity of data generated across the lifecycle of a compound which needs to be linked and interpreted. For example, it is common for a compound developed internally to be transferred between 3-4 sites and up to 30 separate systems prior to commercialization. Each system has valuable information that helps to drive key design and development decisions. As more data and information is gathered across these source systems, it becomes exponentially more difficult to manage the isolated data streams resulting in very high barriers to effectively use the information to create valuable knowledge.

To overcome these barriers, we will introduce you to a recipe based framework that is being implemented to facilitate knowledge lifecycle management. The strategy will enable more routine application of multivariate techniques for which case studies will be demonstrated.

(189e) Variable Selection in Multivariate Modeling of Drug Product Formula and Process

Yong Cui[1], Xiling Song[2], King Chuang[1], Cadapakam Venkatramani[1], Sueanne Lee[3], Gregory Gallegos[4], Thirunellai Venkateshwaran[5] and Minli Xie[6], (1)Small Molecule Pharmaceutical Sciences, Genentech, Inc. , South San Francisco, CA, (2)Small Molecule Pharmaceutical Sciences, Genentech, Inc., South San Francisco, CA, (3)Samll Molecule Pharmaceutical Sciences, Genentech, Inc. , South San Francisco, CA, (4)Small Molecule Pharmaceutical Sciences, Genentech Inc., South San Francisco, CA, (5)Small Molecule Pharmaceutical Sciences, Genentech, Inc. , South san francisco, CA, (6)Small Molecule Pharmaceutical Sciences, Genentech, Inc. , south san francisco, CA

Multivariate data analysis methods such as partial least square (PLS) modeling have been increasingly applied to product development, particularly in the quality-by design paradigm. This study applied the PLS modeling to analyze a product development dataset combining a design of experiment study and historical batch data. Attention was paid in particular to the assessment of the importance of predictor variables, and subsequently the variable selection in the PLS modeling. The results indicate that in conducting PLS modeling irrelevant and collinear predictors could be extensively present in the initial model. Therefore, variable selection is an important step in the model optimization. The VIP and coefficient values can be employed to rank the importance of predictors and to help remove irrelevant predictors. To reduce collinear predictors, on the other hand, multiple rounds of PLS modeling on different combinations of predictors may be necessary. To this end, stepwise reduction of predictors based on their VIP/coefficient ranking was introduced and appeared to be an effective approach to identify and remove redundant collinear predictors.

(189f) A Novel QbD Tool for Statistical Data Analysis

Linas Mockus, Purdue University, West Lafayette, IN, José Miguel Laínez, School of Chemical Engineering, Purdue University, West Lafayette, IN and G. V. Reklaitis, Chemical Engineering, Purdue University, West Lafayette, IN

One of cornerstones of Quality by Design (QbD) is Design of Experiments (DoE). However, in addition to DoE, ICH Q8 (R2) recommends an enhanced approach to product and process development which is based on mechanistic first principle models. Such predictive models require parameters such as kinetic constants which are functions of Critical Process Parameters (CPPs) and determined via designed experiments. Therefore, a robust statistical analysis is a key step for the in-depth process understanding. In this paper we describe a novel technique which is both statistically sound and computationally efficient.

Markov chain Monte Carlo (MCMC) approacheshave provided the computational machinery that has brought Bayesian inference applications to a variety of fields. However, the computational burden that they may require for the inference of complex models involving a considerable number of parameters can be prohibitive. As an alternative, variational approximation methods have been proposed. This family of methods translates the inference problem into an optimization problem by substituting the posterior distribution by a functional form. A decomposition strategy has been developed for the variational approximate inference to deal with models described by differential-algebraic equations (Laínez et al. 2012). In this work we further extend this framework to handle hierarchical models whose number of parameters generally increases with the number of "submodels". The relationship between the hyperparameters and "sub-model"parameters is exploited in order to apply a Lagrangian based decomposition so as to reduce the computation expense of the optimization problem.

A pharmaceutical reaction kinetic case study is presented to demonstrate the advantages of the proposed approach against MCMC approaches for hierarchical models.

References

Laínez, J.M., L. Mockus, S. Orcun, G. Blau, and G.V. Reklaitis. "A decomposition strategy for the variational inference of complex systems", submitted to Technometrics (2012).

Acknowledgements

Support from the United States National Science Foundation(Grant NSF-CBET-0941302) is gratefully acknowledged.

(189g) Mixture Component Prediction Using Iterative Optimization Technology

Koji Muteki[1], Daniel O. Blackwood[1], Kyle R. Leeman[1], George L. Reid[1], Yong Zhou[1], Yang A. Liu[1], Brent Maranzano[1], Paul R. Gerst[1], Howard W. Ward[1], Andreas AM. Mühlenfeld[2] and Matthias Danner[2], (1)Pfizer Worldwide Research and Development, Groton, CT, (2)Pfizer GMS, Freiburg, Germany

Spectroscopy techniques such as near infrared (NIR), mid Infrared (MIR) and ultraviolet (UV) are commonly used to monitor various process operations such as blending, mixing and chemical reactions in pharmaceutical development. The calibration required for a quantitative measurement can be a labor-intensive and time-consuming to implement. The objective of this study is to predict/monitor/estimate the mixture component ratio without making calibration standards in advance (e.g., minimum effort) while keeping similar prediction ability as with calibration standards. Reference methods such as HPLC/GC are used just for the validation purpose.

Iterative Optimization Technology (IOT) involves the use of an optimization computation using only pure component spectra collected prior to a batch and mixture component spectra collected during a batch. The proposed approach has been successfully applied to many practical projects (solvent mixtures, chemical reactions, feed-frame process, blending process). It is expected that the use of IOT will enhance the adoption of PAT as a routine analytical tool in project labs, especially for process understanding and early development activities.

(310a) Optimal Design of Transient Microreactor Experiments

Spencer D. Schaber and Paul I. Barton, Chemical Engineering, Massachusetts Institute of Technology, Cambridge, MA

Development of synthetic routes for active pharmaceutical ingredient (API) synthesis is increasingly performed using microreactors, because their use can reduce experimental cost (both time and material use), make hazardous reactions safer, and enable more precise control [1,2]. While microreactor experiments can be designed to find conditions maximizing yield or selectivity by adjusting processing conditions, others can be designed to discriminate between reaction mechanisms and identify rate parameters as precisely as possible. The latter approach allows more flexibility because mechanisms and model parameters (such as kinetic and equilibrium constants) remain relevant for a range of reactor inlet conditions. This is useful during iterations in a full process design, allowing a mathematical model for a whole process to be optimized. With a dynamic process model in hand, the process analytical technology (PAT) and control systems can be designed as well. Model-based optimal experimental design techniques have been used to maximize the information content of experiments, allowing a single mechanism to be selected and its parameters to be estimated precisely, and the results used to scale up to a 500x larger flow reactor [3]. McMullen and Jensen [3] used a sequence of steady-state experiments in a fully-automated microreactor configuration to choose the best reaction mechanism, then fit a preexponential factor and activation energy. It used relatively little experimental time and material due to automation and the small size of the microreactor. Still, ever faster and cheaper process development is desired to scale up and certify processes, and enter the market as soon as possible. Here we present an evolution of the above concept, in which transient experiments are designed rather than steady-state ones. By modeling the behavior of the system at all times instead of only at steady state, any physically-relevant input functions (such as ramp functions for temperature and flow rate) can be used and measurements can be taken at any time (subject to the limitations of the measurement technology). This eliminates the waste of time and material which occurs between steady states in sequences of steady-state experiments, thus allowing more informative experiments to be performed at less cost. We present case studies with a pharmaceutically-relevant reaction to show for a fixed experimental cost, the degree to which transient experiments reduce uncertainty in (i) parameter estimates and (ii) product purity. We also show for a fixed parameter precision, the degree to which experimental resources can be reduced by optimally designing transient experiments.

[1] Roberge, D. M.; Ducry, L.; Bieler, N.; Cretton, P.; Zimmerman, B. Microreactor technology: a revolution for the fine chemical and pharmaceutical industries? Chemical Engineering and Technology 28 (3), 318–323, 2005.

[2] Ahmed-Omer, B.; Brandt, J. C.; Wirth, T. Advanced organic synthesis using microreactor technology. Organic and Biomolecular Chemistry 5, 733–740, 2007.

[3] McMullen, J. P.; Jensen, K. F. Rapid determination of reaction kinetics with an automated microfluidic system. Organic Process Research and Development 15, 398–407, 2011.

(310b) Building Process Robustness and Flexibility Through Implementation of QbD Methodologies

Kevin Sirk[1], Lorrie Berwick[2], Jason Dorwart[2], Rositza Petrova[3], Paul Fernandez[4] and Luke Schenck[5], (1)Chemical Process Development and Commercialization, Merck & Co., Inc., Rahway, NJ, (2)Analytical Development and Commercialization, Merck & Co., Inc., Rahway, NJ, (3)Center for Material Science and Engineering, Merck & Co., Inc., Rahway, NJ, (4)Chemical Process Development and Commerciliazation, Merck & Co., Inc., Rahway, NJ, (5)Chemical Process Development and Commercialization, Merck & Co, Inc., Rahway, NJ

Effective Quality by Design (QbD) achieves a thorough understanding of processes via a systematic set of development activities guided by a comprehensive risk assessment. The output of the thorough process understanding is a well-defined control strategy that achieves consistent delivery of a product's critical quality attributes (CQAs). Elements of this control strategy include raw material specifications, in process testing, and control via design space definition; moving away from the prior methodology of "testing in quality."

This talk focuses on the QbD development work related to a drug substance crystallization process. Factors from the risk assessment were investigated in a multi-factor Design of Experiments (DOE), with design space ranges selected such that the output of the DOE demonstrates the following: the process ensures all drug substance CQAs are achieved, the process is robust and well understood, and the process is flexible to accommodate the supply chain. Responses from the DOE were statistically analyzed to evaluate their impact on drug substance CQAs across the design space. The results will show that all drug substance CQAs were achieved across the entire design space, demonstrating a robust process that delivers quality API.

(310c) Scale-up Considerations for a Continuous Process with QbD in a Manufacturing Setting

Stephen B. Jeffery, Process Engineering, Eli Lilly and Company, Kinsale, Ireland

To be further developed.

A minimum of 150 words is required for the abstract text, and no more than 1500 words may be submitted. Each abstract is allowed to include up to 20 figures.

As part of a new process API introduction we undertake an equipment comparability study to justify equipment selection scale up decisions. This part of the technical transfer process requires that the knowledge developed at smaller scales is understood at the larger scale, in particular that process parameters, the basis of safety and operability issues are adequately considered for each unit operation and risks identified and controlled.

This talk will outline the issues arising when applying this procedure to continuous processes at manufacturing scale at Lilly's Kinsale, Ireland facility, taking QbD principles into account and translating these to a real manufacturing plant.

This talk will outline the issues arising when applying this procedure to continuous processes at manufacturing scale at Lilly's Kinsale, Ireland facility, taking QbD principles into account and translating these to a real manufacturing plant.

(310d) Application of the Discrete Element Method (DEM) to Understand a Powder Blending Process and De-Risk Scale-up

William R. Ketterhagen and Patrick Daugherity, Worldwide Research and Development, Pfizer Inc., Groton, CT

Scale-up of batch manufacturing operations in the pharmaceutical industry often presents technical challenges and various risks. In this work, a case study describing the use of a discrete element method (DEM) model to study a batch blending process is described. This model was developed to (1) gain insight into undesirable blend uniformity results observed during the manufacture of a recent drug product registration batch and (2) to assess the risk associated with scaling-up the process. During execution of the pilot-scale registration batch, the team had observed the presence of a sub-potent region of the blend that was insufficiently mixed in addition to variability in the potency of stratified samples. The model has confirmed that poor blending is to be expected in this particular bin and visually depicts the presence of a "dead zone" where mixing does not occur due to the particular processing conditions used in the operation of this blender. The model also revealed that, by employing certain modifications to the process, a uniform blend could be achieved in this equipment. To assess the risk in scaling-up this process, the blending performance in the commercial scale bin was predicted to be satisfactory. Results from the commercial scale batch confirm that a well-mixed blend is in fact achieved in this equipment. The use of process models such as this enable scientists and engineers to better understand the process and the associated risks which help to enable successful drug product development and scale-up.

(310e) Validation of a Scale-up Model for the Lubrication Blending Process From Material Sparing to Commercial Scale

Joseph Kushner IV[1], Holger Schlack[2], Margit Karle[2] and Deborah Booth-Schindler[2], (1)Drug Product Design, Pfizer Worldwide Research and Development, Groton, CT, (2)Pfizer PPD, Freiburg, Germany

Turbula bottle blenders are often used in material sparing scale experiments during early-stage pharmaceutical product development. However, applying knowledge gained with these blenders to larger scale diffusion mixers is limited by the lack of blending models that are validated with commercial scale equipment. The present investigation attempts to incorporate Turbula bottle blenders and commercial scale blenders into a recently-developed lubrication blending process scale-up model:

$QA/QA_0 = (1-\beta) + \beta*\exp(-\gamma*L*F_{Headspace}*r)$

where QA_0 is initial quality attribute value, β is sensitivity of the QA to lubrication, γ is formulation-specific lubrication rate constant, L is characteristic mixing length scale, $F_{Headspace}$ is the blender headspace fraction, and r is the number of blending cycles.

2:1 blends of microcrystalline cellulose and spray-dried lactose or dibasic calcium phosphate were mixed with 1% magnesium stearate using Turbula bottle blenders and pilot/commercial scale blenders, varying blender volume, V, (30-1250 ml; 100-2000 Liters); $F_{headspace}$(30, 70%); and r (24 to ~190,000 cycles). Tablets were manufactured using a Huxley-Bertram compaction simulator to evaluate the change in tablet tensile strength as a function of the extent of lubrication mixing. Other attributes of the blends/tablets were also evaluated including compressibility, bulk/tapped density, friability, and disintegration.

The validity of the proposed lubrication blending process scale-up model was shown for blenders from 30-ml to 2000-liter (an approximately five order of magnitude change in scale) for both placebo formulations. This validated scale-up model can be used to replace commercial-scale experiments with bench-scale, material-sparing experiments to support QbD lubrication blending process understanding and design space verification.

(310f) Quality by Design Development of a Film Coating Process Resulting in a Dimensionless Design Space

Alan R. Silverman[1], W. Mark Eickhoff[2], Fochive N. Njikam[1] and Manoharan Ramasamy[3], (1)Pharmaceutical Commercialization Technology, Merck & Co., Inc., West Point, PA, (2)Pharmaceutical Commercialization Technology, Merck & Co., Inc, West Point, PA, (3)Process Analytical Technologies, Merck & Co., Inc., West Point, PA

The Quality by Design development of a film coated tablet drug product began with a risk assessment which identified moisture sensitivity as a key product quality attribute risk. Additionally, tablet moisture was viewed as a key measure for film coated tablet elegance attributes. The film coating unit operation, where the drug product was sprayed with an aqueous based coating suspension, was a key focus of this development. A theory was devised that the tablet moisture could be described by overall thermodynamic factors, spray and tablet flux factors, and geometric factors. A design of experiments was executed at both lab and commercial film coating scales. The result of the design of experiments was the identification of the high, medium, and low risk process factors impacting the tablet moisture response via a statistical model in support of this theory. The design of experiments indicated the scale independence of the moisture response over the range of study. This allowed the description of the film coating process design space with a collection of dimensionless groups in the product New Drug Application. This film coating control strategy was accepted by regulatory agencies with the approval of the product and it has been implemented at the commercial site, first with process validation and then with subsequent commercial production.

(375a) A Regulatory Perspective On the Current and Future State of Pharmaceutical Manufacturing

Christine Moore, CDER/OPS/ONDQA, FDA, Silver Spring, MD

For almost one decade, the US FDA has been encouraging pharmaceutical manufacturers to modernize their manufacturing processes, improve their process and product understanding, and adopt state of the art quality management systems. These efforts have spurred great advances in science and risk based approach for pharmaceutical development and manufacturing such as Quality by Design (QbD) and Process Analytical Technology (PAT). QbD has become a development standard in many innovator pharmaceutical companies and increasingly is being adopted by biotech and generic firms.

This presentation will cover a brief background in FDA's quality initiatives and QbD principles as outlined in International Conference on Harmonization (ICH) guidances, ICH Q8(R) "Pharmaceutical Development," ICH Q9 on "Quality Risk Management," and ICH Q10 on "Pharmaceutical Quality Systems." Examples will be presented of advanced pharmaceutical manufacturing currently being practiced. Finally, future directions in advanced pharmaceutical manufacturing will be discussed, including multivariate approaches to batch "process signatures" and continuous manufacturing.

(375d) Achieving Manufacturing Flexibility and Product Quality Through Dynamic Design Spaces

James K. Drennen III, Mylan School of Pharmacy, Duquesne University, Pittsburgh, PA

Flexible manufacturing through dynamic design spaces? Mechanistic models vs. empirical models? Quality by design provides opportunity for the pharmaceutical industry to apply multivariate modeling methods in defining design spaces. Modern approaches to process validation have led to interesting questions regarding the definition of design spaces. Is a design space best defined as the multivariate model that relates critical process parameters to critical quality attributes? What combination of mechanistic and empirical modeling will provide an optimum design space while meeting current regulatory philosophy? What are appropriate methods for managing model (design space) updates? How are models or design spaces best verified? What is the effect of scale on models that define design spaces? What is a "significant" change in design space? What are appropriate tolerances on a design space? These and other current questions regarding design spaces will be considered by the author in discussing modern approaches to achieving manufacturing flexibility and optimizing product quality.

(375e) Modeling and Simulation: Historical Parallels Between the Pharmaceutical and Petrochemical Sectors

Salvador García-Muñoz, Process Modeling & Engineering Technology, Pfizer Worldwide Research & Development, Groton, CT

A retrospective analysis on the technical evolution and cultural acceptance of modeling and simulation technology in the petrochemical and polymer sector is presented. Although the sophisitication of the tools difers due to the natural progression of technology, the cultural barriers and challenges observed decades ago in the petrochemical sector are not too different from those seen today in pharma. Such an analysis is presented to encourage pharmaceutical industrials to join efforts in pre-competitive areas to leverage the combined pool of modeling talent and experience for the benefit of the sector, and to work with vendors and academics to close the gap between theoretical developments and practical implementations.

(375g) Panel Discussion (1hr slot)

Christopher L. Burcham, Chemical Product Research and Development; Particle Design Laboratory, Eli Lilly and Company, Indianapolis, IN and **Huiquan Wu**, Division of Product Quality Research, HFD-940, FDA, Silver Spring, MD

Round Table Invited session on QbD: QbD filings As FDA continues to facilitate the modernization of regulatory processes related to pharmaceutical products in the 21st Century, Quality-by-Design (QbD) approaches for some of the pharmaceutical process/product development and manufacturing have been discussed intensively over the past few years in various forums. Some real-world case studies have been reported in the public literature. However, a risk-based and science-based approach for QbD filling based on the regulatory guidance and guidelines are barely shared and discussed in the public domain. As a first step, this round table discussion session is intended to address this information gap to the possible extent. Inputs from recognized experts and experienced practitioners in the community across academia, industry, and regulatory agencies in the world will be sought to identify commonality among QbD filing. An in-depth panel discussion following the presentations should stimulate innovative research and development in the QbD area, and facilitate the QbD implementation in the pharmaceutical sector.

The session chair/co-chair will lead the discussion. All of the invited speakers will be panelists. Active participation from the session attendees will be encouraged.

Process Understanding Tools in PAT Implementations

João Machado, Tiago Robalo, Pedro Felizardo
4Tune Engineering Ltd - Technology Services & Products, Lisbon, Portugal

Abstract

For the past years the pharmaceutical industry has been trying to increase its process understanding in order to keep pace with regulatory demands and also to increase its competitiveness by being able to find improvement opportunities and increase efficiency. This is also a requirement from the new paradigm of Quality by Design.

For this increase in process understanding demand, the industry has started to collect increasingly amounts of new data from the process and the data itself is now of different types, particularly with the introduction of PAT tools. These tools that are able to collect complex data (NIRS, 2D-Fluorescence, dielectric spectroscopy, etc.) pose challenges for the data organization, structuring, accessibility, prioritization and knowledge extraction. In fact, without these challenges solved, the amount and type of data collected can reveal to be a considerable investment and irrelevant for practical use.

If data is not systematically and automatically organized (presuming that it is being automatically gathered), making sense of it can become an unreasonably time-consuming activity. Diagnostic and troubleshooting efforts are lost in the data matrix complexity and energy is spent in the organization of the data in the context of the production process. Also, this lack of structure and organization also poses barriers to the implementation of prediction and MSPC models across the entire cascade of process steps (from raw-materials to end product). To perform true whole-process analysis across the entire process and process life-cycle in the lights of the new guidelines and regulatory demands, this issue needs to be solve with new tools that enable the shift of energy from data organization to the more useful process understanding and process knowledge sphere with a true system engineering environment.

Therefore a new tool that provides a solution to the industry to tackle these problems is needed. This tool should allow the organization and storage (with proper structure) of different types of data: Process data (temperatures, pressures, flow rates, etc.), PAT fingerprint data (NIR. MIR, 2D-Fluorescence, etc.), LIMS data and in-line monitoring data. These data have to be added in real-time or be a part of an historical database of the process enabling fast monitoring and diagnosis of the process or historical analysis of previous runs. The necessary graphical instruments are also required in this tool to enable quick and intuitive visualization of the data according to its type (single plots, landscapes, etc.).

Comparison of data should also be possible in this graphical interface that takes the previous automatic organization of data and allows easy evaluation of different data types, from different process steps (ex: unit operations) at different time steps or from different runs. This should prove very useful when performing root-cause analysis and assessing cause & effect on the process.

Many of the data being gathered from pharmaceutical production process is used (or could be used) to predict production quality, control the process, diagnose performance, assist in real time release, etc. This is especially true since the introduction of PAT tools that can extract increasingly more informative data from the process. By applying modeling techniques to these data (PAT data, process data, etc.) more process understanding and control can be implemented. For these reasons, this tool should support an easy integration of inline prediction models, calibrations, MSPC, etc. from third-party software. In fact, by having this feature, the user has the chance to improve its models in accuracy and robustness by taking advantage of the organization and accessibility of all the data that ultimately leads to a greater process understanding.

The importance of fast sharing of information and findings is an increasing necessity in the industry when knowledge-based decisions often have to be taken with haste but also truthful. In this context, a new tool has to support mechanisms to share the information found (through normal channels such as email, PDF, etc.) and also to store it (with systems like dropbox, database storage or similar).

In order to exemplify the above-mentioned needs and the advantages of such tool in the pharmaceutical industry, it is intended to present iSEE™ (Systems Engineering Environment interface) in the context of a case study with data from a bio-production pharmaceutical process.

2012 AIChE Annual Meeting QbD Topical Conference
(Pittsburgh, PA, Oct 28-Nov 2 2012)
Session 401 "**Application of PAT in the Manufacturing Setting**"
Paper 401c

An Integrated Process Analytical Technology (PAT) Approach to Examine the Effect of Temperature on Nucleation Kinetics of a Dynamic Pharmaceutical Co-precipitation Process

Huiquan Wu*, Maury White, Ryan Foringer, and Mansoor Khan
Division of Product Quality Research (DPQR, HFD-940), OTR/OPS/CDER/FDA
FDA White Oak Research Center, 10903 New Hampshire Ave
Silver Spring, MD 20993

Abstract: Process Analytical Technology (PAT) has become an essential part for pharmaceutical process and product understanding, real-time process monitoring and control during the last several years. Previously, we reported the development of a novel real-time PAT-based approach to measure the nucleation induction time and elucidate the nucleation and growth mechanisms of a dynamic pharmaceutical co-precipitation process at 25^0C. The goal of this work is to understand the effect of temperature (15^0C, 25^0C, 35^0C) on nucleation induction time for a dynamic pharmaceutical co-precipitation process (naproxen-Eudragit L100-alcohol-water) at various ratios of drug/polymer. Real-time inline process monitoring and nucleation induction time (t_{ind}) measurement have been accomplished using FBRM and PVM simultaneously. It was shown that: (1) the plots of t_{ind} vs. the ratio of naproxen to polymer (Q) display two distinct linear segments; (2) the plots of $\ln(t_{ind})$ vs. $(\ln S)^{-2}$ (S is the maximum allowable super-saturation level) display two distinct linear segments which agree well with the classic nucleation theory (CNT); (3) Quantitative relationships based on CNT among the kinetic parameter A, dimensionless thermodynamic parameter B, and temperature (T) suggest that the nucleation and growth process is followed by the interface transfer mechanism. Therefore, at low S level, the nucleation is governed by primary heterogonous nucleation; while at high S level it is governed by primary homogeneous nucleation; after nucleation, the growth is controlled by the interface transfer mechanism. This work demonstrated the utility of real time process PAT monitoring on process understanding and process kinetic mechanism elucidating, thus could enhance the science-based pharmaceutical regulation in the Quality-by-Design (QbD) domain.

Key words: PAT, QbD, Nucleation and Growth Kinetics, Dynamic Co-precipitation Process, Nucleation and Precipitation Induction Time, Dimensionless Thermodyanmic Parameter and Kinetic Parameter

*Corresponding author: Huiquan Wu (Telephone: 001-301-796-0095; Fax: 001-301-796-9816; Email: Huiquan.wu@fda.hhs.gov)

Disclaimer: The views and opinions expressed in this work are only of the authors, and do not necessarily reflect the policy or statement of the US FDA.

1. PAT approach to examine the nucleation phenomena and crystallization onsets

Previously, an integrated PAT (FDA, 2004) approach based on real time Focused Beam Reflectance Measurement (FBRM), real time Process Near-infrared (NIR) Spectroscopy, and process chemometrics had been developed for measuring and validating the co-precipitation nucleation induction time, and elucidating the nucleation and growth mechanisms of a 4-component co-precipitation process at 25 ^{0}C (Wu and Khan, 2011).

Several experimental techniques have been used for examining the nucleation phenomena and crystallization onsets, such as polarized light microscopy (Varon et al, 2010), Focused Beam Reflectance Measurement (FBRM) and Particle Vision Microscopy (PVM) (Wu and Khan, 2011). Nucleation induction time (Söhnel and Mullin, 1988) (t_{ind}) is a quantitative parameter for characterizing the nucleation tendency for solution crystallization at isothermal condition. It is the time elapsed between super-saturation creation and new phase detection in a system. It is a function of the initial super-saturation and temperature. However, it is not a fundamental property of a system because the new phase detection is open to a variety of interpretations, such as the first appearance of crystals or the onset of a change of solution property. There are a number of experimental techniques used for new phase detection, such as differential scanning calorimetry (Yosjihashi et al., 2010), conductivity measurement (Van Der Leeden et al., 1992), turbidity measurement (Guo et al., 2006), intensity of transmitted or scattered light method (Wakita and Masuda, 1983), and combined process NIR and turbidity measurement (Wu and Khan, 2010).

From the transport process perspective, it is important to recognize that the nucleation and growth process involves multiple steps, including nucleation initiation, bulk diffusion, and interface transfer (between the nucleus and substrate). Depending on the process conditions and the crystallization system being studied, one or a combination of these could be the rate-limiting step for the nucleation and growth process. In theory, the impacts of critical process parameters (such as temperature and super-saturation level) on nucleation and growth kinetics are highly complicated. Based on our previous Design of Experiments (DOE) and ANOVA results (Wu et al.,2011), temperature is the most influential process variable with respect to its impact on several response variables for the co-precipitation process, including derived process rates during transition periods, mode and peak frequency at the end of steady state. In this work, the effect of temperature on nucleation kinetics and growth kinetics is characterized by the measurement and analysis of the nucleation induction time via an integrated real-time online PAT monitoring scheme, under the framework of Classical Nucleation Theory (CNT).

2. Experimental

Naproxen USP (lot No: NPX 368) was obtained from Albemarle Corporation (Orangeburg, SC). Eudragit L 100 (lot No: 1221203048) was obtained from Röhm America Inc. (Somerset, NJ). Solvent reagent alcohol (HPLC grade, lot No: 053546) was purchased from Fisher Scientific (Fair Lawn, NJ). Non-solvent DI water was obtained from a FDA in-house facility (Silver Spring, MD) and was kept refrigerated at 3^{0}C prior to use. All chemicals, solvent, and non-solvent were used without any further processing or purification prior to the use for this experimental work.

Experimental setup is similar to previous work, except that both FBRM and PVM are simultaneously monitoring the dynamic co-precipitation process in this work. Both FBRM and PVM data were acquired every two seconds.

To quantify the effect of temperature on nucleation and growth kinetics mechanisms for this pharmaceutical co-precipitation system, other process parameters were fixed throughout the study: stirring rate 200 rpm; DI water addition rate 1.8 ml/sec. Various ratios of API/Polymer (Naproxen)/Eudragit L100) were used for the study, although it was fixed for each individual experiment. Nucleation induction time was measured for each co-precipitation system based on real-time inline PAT monitoring results. As discussed in our previous work (Wu and Khan, 2011), the nucleation induction time measurement for this particular dynamic co-precipitation system has been quantitatively validated by an independent method, which is based on process trajectory and real-time online Near-infrared (NIR) spectroscopic monitoring results. In this work, the nucleation induction time measurement based on FBRM chord-length-distribution (CLD) was qualitatively validated by real time online PVM measurement results. Theoretical analysis of those nucleation induction times and corresponding super-saturation ratios was carried out based on the classical nucleation theory (CNT), as discussed below.

3. Classical Nucleation Theory (CNT)

Based on the classical primary nucleation theory (Nývlt 1968; Sőhnel and Mullin, 1988; Mullin 1972; Kashchiev et al., 1991), we have

$$\ln(t_{ind}) = \ln(\frac{N_c}{VA_{\hom,J}}) + \frac{16\pi\gamma^3 v^2}{3(kT)^3} \cdot \frac{1}{(\ln S)^2} \quad \text{for HON process;} \qquad (3.1)$$

$$\ln(t_{ind}) = \ln(\frac{N_c}{VA_{het,J}}) + \frac{16\pi\gamma^3 v^2}{3(kT)^3} \cdot f(\theta) \cdot \frac{1}{(\ln S)^2} \quad \text{for HEN process.} \qquad (3.2)$$

Eq. (3.1) and Eq.(3.2) suggests that, at a given temperature, the plot of $\ln(t_{ind})$ vs. $(\ln S)^{-2}$ should give a straight line. For a HON process, the slope and intercept are:

$$B_{\hom,t_{ind}} = \frac{16\pi\gamma^3 v^2}{3(kT)^3} \qquad (3.3)$$

$$A_{\hom,t_{ind}} = \ln(\frac{N_c}{VA_{\hom,J}}) \qquad (3.4)$$

For a HEN process, the slope and intercept are:

$$B_{het,t_{ind}} = \frac{16\pi\gamma^3 v^2}{3(kT)^3} \cdot f(\theta) \qquad (3.5)$$

$$A_{het,t_{ind}} = \ln(\frac{N_c}{VA_{het,J}}) \qquad (3.6)$$

For HON mechanism,

(1) If growth is volume-diffusion controlled, we have

$$A_{\hom,t_{ind}} = \ln(\frac{N_c}{V}) + \ln(\frac{1}{A_{\hom,j}}) = A_{\hom}^{diff} - 0.5\ln T \quad (3.7)$$

(2) If growth is interface transfer controlled, we have

$$A_{\hom,t_{ind}} = \ln(\frac{N_c}{V}) + \ln(\frac{1}{A_{\hom,j}}) = A_{\hom}^{surf} + 0.5\ln T \quad (3.8)$$

For HEN mechanism,

(1) If growth is volume-diffusion controlled, we have

$$A_{het,t_{ind}} = \ln(\frac{N_c}{V}) + \ln(\frac{1}{A_{het,j}}) = A_{het}^{diff} - 0.5\ln T \quad (3.9)$$

(2) If growth is interface transfer controlled, we have

$$A_{het,t_{ind}} = \ln(\frac{N_c}{V}) + \ln(\frac{1}{A_{het,j}}) = A_{het}^{surf} + 0.5\ln T \quad (3.10)$$

Therefore, for both HON and HEN mechanisms, the slope of $A_{\hom,t_{ind}}$ vs. $\ln T$ and the slope of $A_{het,t_{ind}}$ vs. $\ln T$ will be negative if the growth is volume-diffusion controlled, and positive if the growth is interface-transfer controlled.

4. Results and Discussion

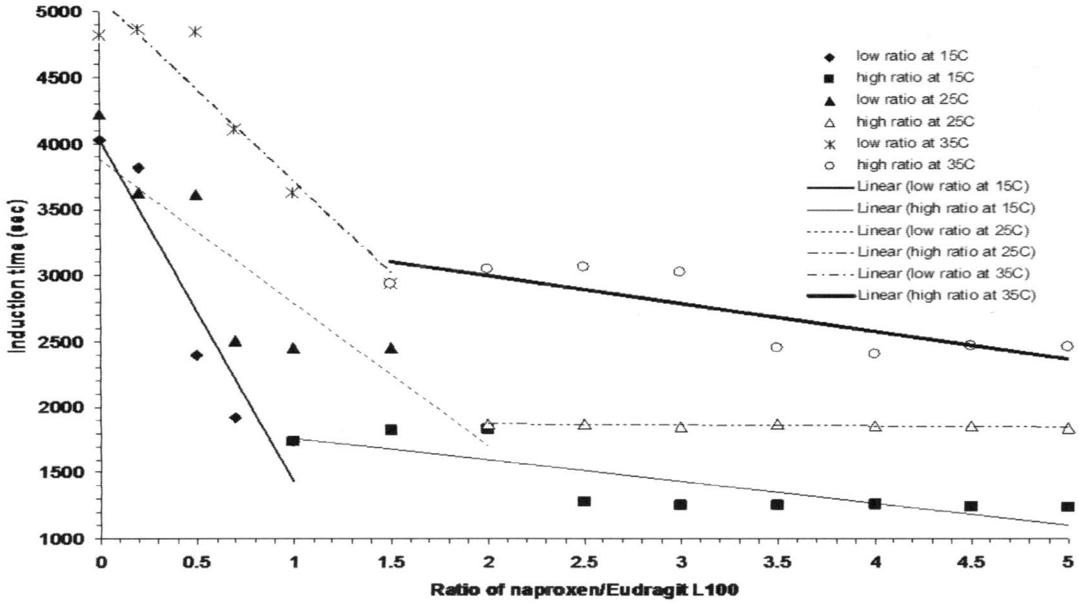

Figure 1. Plots and linear regression lines of nucleation induction time vs. drug/polymer ratio at15^{0}C, 25^{0}C, and 35^{0}C.

Figure 1 shows that there are two distinct segments for each temperature: at low drug/polymer ratio, the induction time decreases quickly with the increase of drug/polymer ratio; after the drug/polymer ratio reaches a threshold value which changes with the temperature, the decreasing rate of induction time vs. drug/polymer ratio becomes much smaller. Eventually, the induction time reaches a plateau value specific to the temperature.

Figure 2. Nucleation and/or precipitation induction and growth phenomena monitored in real time by FBRM and PVM simultaneously. PVM images and FBRM CLD profiles at 15 ^{0}C, with various *Q*.

(2a) *Q*=1.0 at time point of t_{ind}

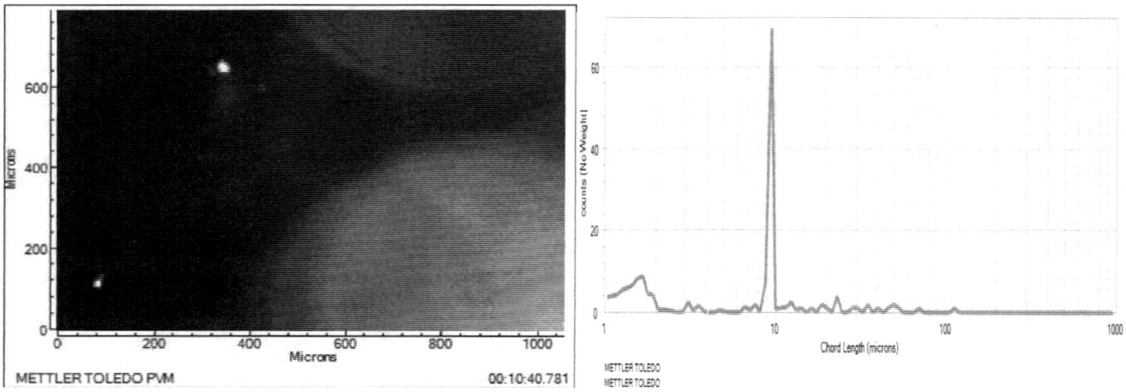

(2b) *Q*=1.0 at time point of t_{ind} +10 minutes

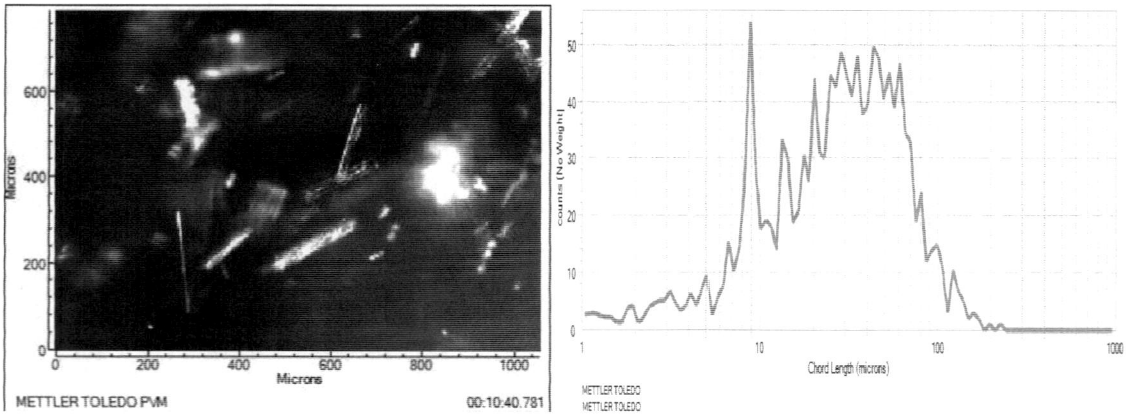

(2c) *Q*=5.0 at time point of t_{ind}

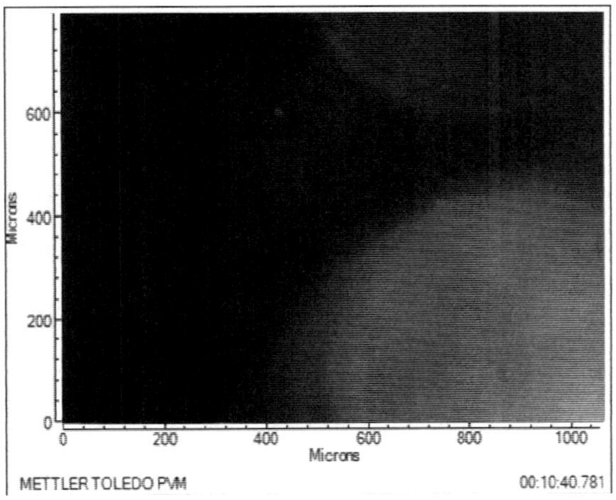

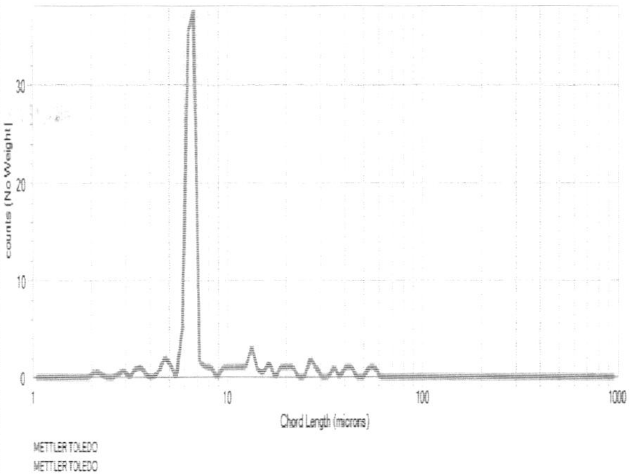

(2d) Q=5.0 at time point of t_{ind} +10 minutes

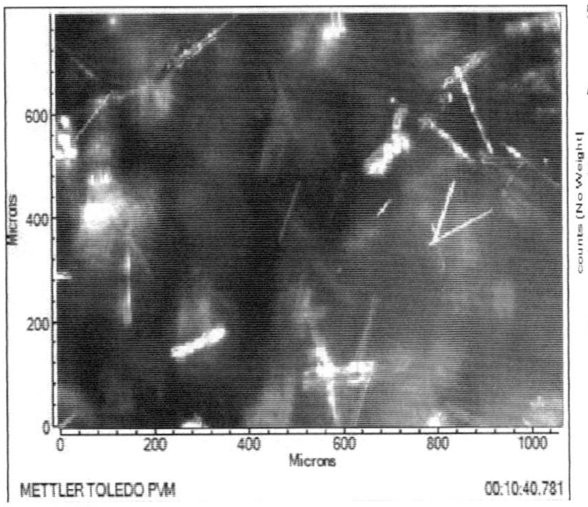

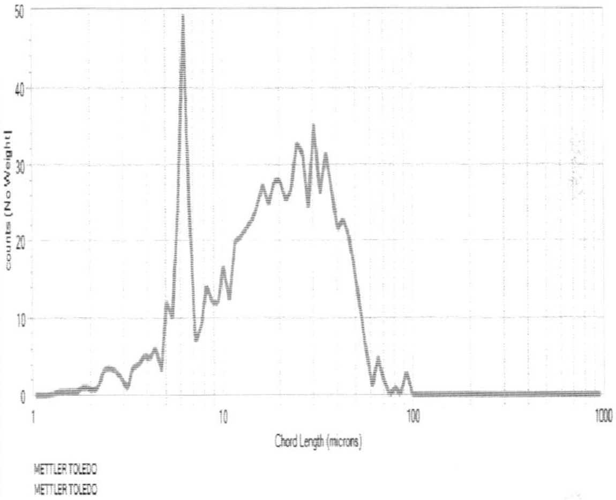

At time point of t_{ind}, PVM pictures show several small particles are detected, while FBRM shows that the counts/s is in the range of magnitude of 10. At time point of t_{ind} +10 minutes, the increased number of crystals and increased dimensions of particles are evident by PVM pictures, while the increased dimensions of particles and crystals are also evidenced by FBRM CLD.

Figure 3. Nucleation induction time as a function of super-saturation. Plots and regression lines of $ln(t_{ind})$ vs. $(lnS)^{-2}$ at15^{0}C.

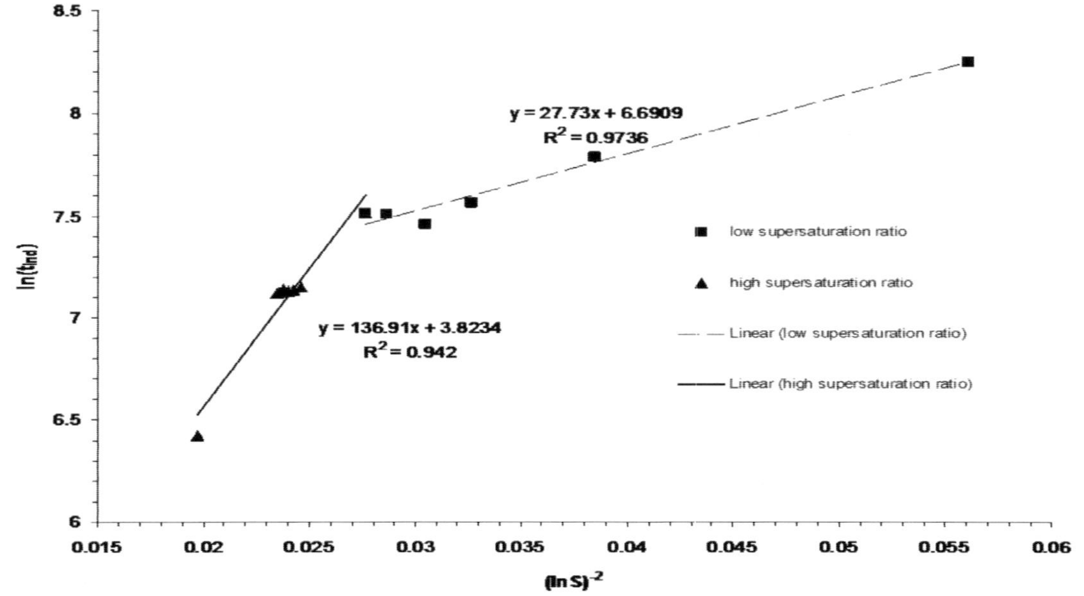

The different nucleation regimes are highlighted and the corresponding parameters are indicated. At low S level, the nucleation is governed by heterogeneous nucleation mechanism; at low S level, the nucleation is governed by homogeneous nucleation mechanism.

Figure 4a Plots and regression lines of the dimensionless thermodynamic parameter *B* vs. T^{-3} at both low and high S value.

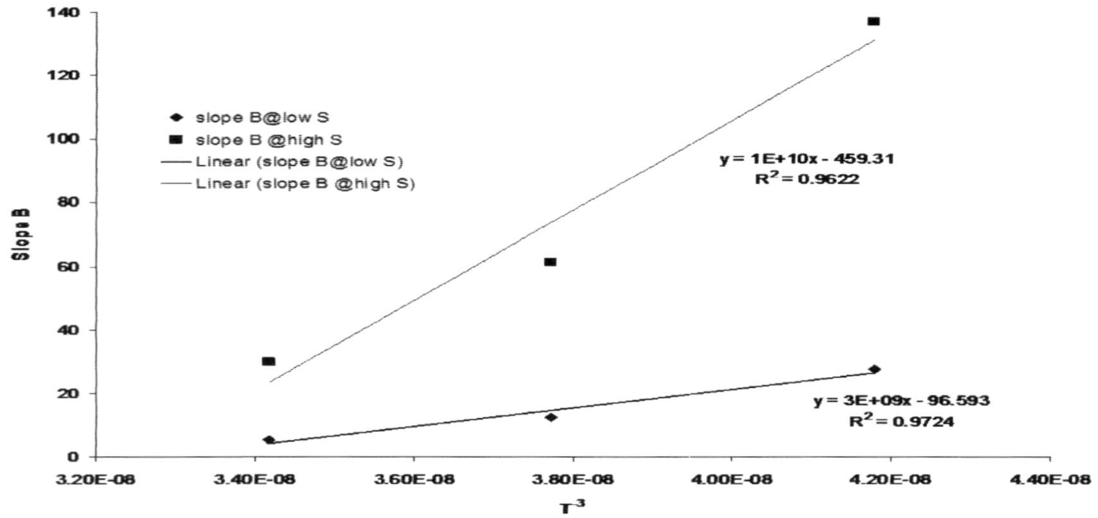

At low S segment, the HEN mechanism prevails; at high D segment, the HON mechanism is domaint.

Figure 4b Plots and regression lines of kinetic parameter *A* vs. ln*T* at both low and high S value.

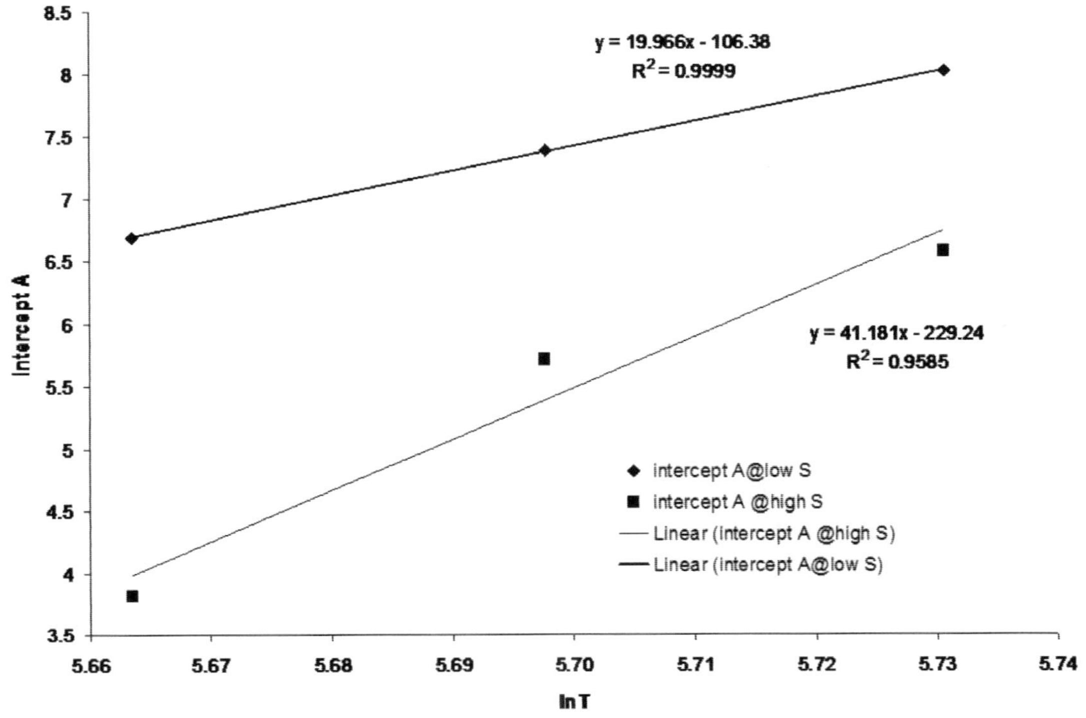

The intercept A changes with lnT in a linear fashion with a positive slope. This suggests that the growth is controlled by the interface transfer mechanism rather than the volume-diffusion mechanism.

5. Conclusions:

An integrated Process Analytical Technology (PAT) approach has been established to examine the effect of temperature on nucleation and growth kinetics of a 4-component dynamic pharmaceutical co-precipitation process. It reveals that at low S level, the nucleation is governed by primary heterogonous nucleation; while at high S level it is governed by primary homogeneous nucleation; after nucleation, the growth is controlled by the interface transfer mechanism.

Acknowledgements:
H.Wu sincerely acknowledge the technical guidance and supports from Dr. Vincent Vilker, Director, Office of Testing and Research (OTR), CDER, FDA (Silver Spring, MD), and Dr. San Kiang at BMS Research Institute (New Brunswick, NJ). The following persons and organizations are thankful: Mr. John Casamento from CDRH, FDA (Silver Spring, MD), Mr. Eric Dycus, Ms. Vaso Vlachos, Mr. Ben Smith from Mettler-Toledo AutoChem, Inc. (Columbia,MD), Mr. Steve Ware and Mr. Brendan Simon from ChemGlass Life Science (Vineland, NJ) for their assistance on instrumentations; Röhm America Inc. (Somerset, NJ) for technical assistance.

References:

1. FDA, 2004. "Guidance for Industry. PAT—A Framework for Innovative Pharmaceutical Development, Manufacturing, and Quality Assurance". Available at: http://www.fda.gov/cder/guidance/6419fnl.pdf (accessed on 10/01/2012)
2. Guo, Z, Jones, A.G., Li, N., 2006. Interpretation of the ultrasonic effect on induction time during $BaSO_4$ homogeneous nucleation by a cluster coagulation model. J. Colloid and Interface Science 297:190-198.
3. Kashchiev, D., Verdoes, D., and van Rosmalen, G.M., 1991. Induction time and metastability limit in new phase formation. J. Crsyatl Growth 110:373-380
4. Mullin, J.W. Crystallization. 2nd Edition. Butterworth-Heinemann Ltd., London, 1972, p.573.
5. Nývlt, J., 1968. Kinetics of nucleation in solutions. J. Cryst Growth 3,4:377-383.
6. Söhnel, O. and Mullin, J.W.1988. Interpretation of Crystallization Induction Periods. J. Colloid and Interface Science 123: 43-50.
7. Van Der Leeden, M.C., Kashchiev, D., and Van Rosmalen, G.M., 1992. Precipitation of Barium Sulfate: Induction Time and the Effect of an Additive on Nucleation and Growth. J. Colloid and Interface Science 152:338-350.
8. Varon, V., Bhugra, C., and Pikal, M.J., 2010. Prediction of Crystallization in Amorphous Pharmaceutical Systems: Phenobarbital, Nifedipine/PVP, and Phenobarbital/PVP. J. Pharm. Sci. DOI 10.1002/jps.22232.
9. Wakita, M.I.H. and Masuda, I., 1983. Analysis of precipitation processes of bis(dimethylglyoximato)Ni(II) and related complexes. J. Crystal Growth 61:377-382.
10. Wu, H. and Khan, M.A., 2010. Quality-by-Design (QbD): An Integrated Process Analytical Technology (PAT) Approach for Real-time Monitoring and Mapping the State of a Pharmaceutical Co-precipitation Process. Journal of Pharmaceutical Sciences 99(3): 1516-1534.
11. Wu, H. and Khan, M.A., 2011. Quality-by-Design (QbD): An Integrated Process Analytical Technology (PAT) Approach to Determine the Nucleation and Growth Mechanisms during a Dynamic Pharmaceutical Co-precipitation Process. Journal of Pharmaceutical Sciences 100(5): 1969-1986
12. Wu, H, White,M., and Khan, M.A., 2011. Quality-by-Design (QbD): An Integrated Process Analytical Technology (PAT) Approach for a Dynamic Pharmaceutical Co-precipitation Process Characterization and Process Design Space Development. *International Journal of Pharmaceutics*, 405:63-78.
13. Yosjihashi, Y., Yonemochi, E., Maeda, Y., Terada, K., 2010. Prediction of the induction period of crystallization of naproxen in solid dispersion using differential scanning calorimetry. J. Therm. Anal. Calorim. 99:15-19.

(401d) Making PAT an Integral and Sustained Part of Manufacturing Operations

John O'Reilly, Roche, Clarecastle, Ireland

To be further developed.

A minimum of 150 words is required for the abstract text, and no more than 1500 words may be submitted. Each abstract is allowed to include up to 20 figures.

Modern process analytical technology (PAT) can promise many innovative solutions in the manufacture of API. However, to implement such technologies in a sustainable way is often the greatest challenge.

This talk looks at some of the current PAT projects at Roche Clarecastle and in particular the innovative strategy we are developing to make PAT an integral and sustained part of our manufacturing operations.

The concepts of in-process data samples (IPDS) and PAT-QC are introduced. The use of PAT as a continuous real-time process verification/validation tool is also discussed with reference to the latest FDA process validation guideline.

The concepts of in-process data samples (IPDS) and PAT-QC are introduced. The use of PAT as a continuous real-time process verification/validation tool is also discussed with reference to the latest FDA process validation guideline.

(401e) Design and Implementation of An Efficient Control System in a Continuous Pharmaceutical Manufacturing Process Via Roller Compaction

Ravendra Singh, Marianthi Ierapetritou and Rohit Ramachandran, Department of Chemical and Biochemical Engineering, Rutgers, The State University of New Jersey, Piscataway, NJ

Pharmaceutical industries are facing enormous challenges to achieve the predefined end product quality consistently as well as to satisfy the regulatory constraints, flexible market demands, operational complexities and economical limitations [1]. A novel manufacturing strategy based on continuous processing integrated with online monitoring tools coupled with efficient automatic feedback control system is therefore highly desired [2]. However, because of different level of complexities involved (e.g. solid handling, irregular flow, multi-variate interactions, difficulty in real time measurement) in addition to the requirement of very precise product quality, the design and implementation of an efficient control system in the pharmaceutical manufacturing involving the solid dosage form is still a challenging task. The process involving the solid flow is not as simple to control as the process involving the liquid/gas flow therefore the conventional control strategy may not be efficient, although it could be adapted. The design and implementation of the process control system involves the identification of critical control variables, coupling of the control variables with suitable actuators, selection of suitable monitoring tools, selection of control strategy followed by controller parameters tuning, model-based closed-loop performance evaluation and finally implementation to the manufacturing plant through the available hardware (e.g. Delta V) and control interface (e.g. OPC).

In the work reported here, an efficient control system for an integrated continuous pharmaceutical tablet manufacturing process via roller compaction has been designed. The continuous manufacturing process consists of the following unit operations: blending, roller compaction, milling, tablet compression and coating. The roller compactor has been used as a means of dry granulation to avoid any API water sensitive issues involved. The designed control system consists of four cascade control loops and three single control loops giving 36 tuning parameters. An effective controller parameter tuning strategy involving ITAE methods coupled with an optimization strategy has also been proposed. The designed control system has been implemented in a first principle model-based flowsheet simulated in gPROMS (Process System Enterprise) to verify its performance and is being implemented in a continuous pilot plot at Rutgers University. The systematic methodology for pharmaceutical process control system design as well as supporting process models and knowledge/data base has also been developed.

The objective of this presentation is two-fold: first to highlight the design of control system for a continuous tablet manufacturing process with roller compaction, and second to demonstrate the performance of the designed control system through the first principle model as well as in a pilot plant.

References

[1] Singh, R., Gernaey, K. V., Gani, R., (2010). ICAS-PAT: A Software for Design, Analysis & Validation of PAT Systems. Computers & Chemical Engineering, 34(7), 1108-1136.

[2] Ramachandran, R., Arjunan, J., Chaudhury, A, Ierapetritou, M. (2012). Model-Based Control Loop Performance Assessment of a Continuous Direct Compaction Pharmaceutical Processes. J. Pharm. Innov., 6(3), 249-263.

(401f) Near Infrared Spectroscopy (NIRS): An Enabling Tool for Pharmaceutical Quality by Design (QbD)

K. Geoffrey Wu, Suhas Patankar, Lane V. Christensen, Robert A. Lionberger and Lawrence X. Yu, Office of Generic Drugs, Food and Drug Administration, Rockville, MD

Near Infrared Spectroscopy (NIRS): An Enabling Tool for Pharmaceutical Quality by Design (QbD)

K. Geoffrey Wu, Suhas Patankar, Lane V. Christensen, Robert A. Lionberger, Lawrence X. Yu

Office of Generic Drugs, Food and Drug Administration, Rockville, MD 20855

This presentation will provide an overview of the application of NIRS within the paradigm of pharmaceutical QbD and elaborate on the current thinking of the Office of Generic Drugs (OGD) regarding the deployment of such a useful application of process analytical technology (PAT).

Applying various types of NIRS methods for pharmaceutical development has been proposed by the generic industry to OGD. Case studies will be given to demonstrate some of the major technical issues observed during the review of these proposals. Such issues may be related to the NIR method development and validation, model maintenance, and parallel testing. Points to consider during the model transfer (e.g., from R&D to manufacturing phase) will also be discussed.

OGD encourages the use of NIRS and considers it as an important tool to facilitate the implementation of pharmaceutical QbD. From OGD's perspective, possible regulatory pathways to file the use of NIRS will be presented.

(445a) Economic Plantwide Control Design Procedure

Rahul Jagtap, Chemical Engineering, Indian Institute of Technology Kanpur, Kanpur, India, Nitin Kaistha, Chemical Engineering, IIT Kanpur, Kanpur, India and Sigurd Skogestad, Department of Chemical Engineering, Norwegian University of Science and Technology, Trondheim, Norway

A general systematic top-down procedure for an effective plantwide control system design for economically optimal operation of a process over a large throughput range is synthesized. As throughput is increased, operating constraints become active one by one at the optimum steady state till maximum plant throughput where maximum number of constraints is active. The crucial step in the systematic procedure is to design a control system for economic process operation at maximum throughput. In step 1 of the procedure control loops for tight control of active constraints and economically dominant CVs are implemented, then in Step 2 inventory control loops are set followed. Step 3 involves control loop configurations for throughput manipulation so that the setpoints that becomes unconstrained at lower throughput can take up additional close-by economic CV control for throughput manipulation. Step 4 involves control structure reconfiguration to eliminate fragile / unconventional inventory control loops resulting from above 3 steps. The procedure is applied on a realistic chemical process.

(445b) Systematic Synthesis of Augmented Biomass-to-Liquid Fuel Processes

Dharik S. Mallapragada[1], Mohit Tawarmalani[2], Fabio H. Ribeiro[1], W. Nicholas Delgass[1] and Rakesh Agrawal[1], (1)School of Chemical Engineering, Purdue University, West Lafayette, IN, (2)Krannert School of Management, Purdue University, West Lafayette, IN

Liquid fuels from biomass potentially offer a sustainable option for supplying high volumetric energy density fuel for the transportation sector in a future solar energy driven economy. However, there are limits on the amount of sustainably available (SA) biomass that can be recovered annually without using additional agricultural land. Since the continued use of liquid fuel for transportation seems likely, conversion processes that maximize the liquid fuel output from the limited SA biomass resource are worthy of exploration.

Here, we present systematic synthesis of augmented biomass-to-liquid fuel processes, which maximize biomass carbon conversion to liquid fuel by using supplemental energy derived from sunlight in the form of H_2, heat, and electricity. While standalone biomass-to-liquid fuel processes are limited to <50% biomass carbon recovered as liquid fuel, augmented processes overcome this barrier by utilizing biomass as a carbon source combined with supplemental solar energy use. The emphasis is on identifying augmented processes that require the least amount of supplemental solar energy input corresponding to a given biomass carbon recovery.

We focus on two promising biomass to liquid fuel thermochemical approaches of heat-assisted biomass gasification followed by fuel synthesis and biomass fast-hydropyrolysis/hydrodeoxygenation (HDO). A superstructure of all possible process configurations for the aforementioned thermochemical routes is developed. Subsequently, an equation-oriented nonlinear optimization model is solved to identify the optimal process configuration for different target carbon recovery levels. The novel features of the process synthesis include: 1) a simplified fast-hydropyrolysis/HDO model derived from experimental literature, 2) simultaneous process heat, power and mass integration while allowing for co-generation and 3) flexibility to design for different carbon recovery levels. The presentation will also briefly discuss the progress in developing a theoretical guarantee of the global optimality of the solutions presented.

(445c) Process Synthesis of Hybrid Coal, Biomass, and Natural Gas to Liquids (CBGTL) Via Fischer-Tropsch Synthesis, ZSM-5 Catalytic Conversion, Methanol Synthesis, Methanol-to-Gasoline, and Methanol-to-Olefins/Distillate Technologies

Richard C. Baliban, **Josephine A. Elia**, Vern W. Weekman Jr. and Christodoulos A. Floudas, Chemical and Biological Engineering, Princeton University, Princeton, NJ

Thermochemical conversion of biomass or coal to liquid hydrocarbons can proceed via syngas intermediate, followed by either the Fischer-Tropsch (FT) process or via methanol intermediate. Thermochemical conversion of natural gas generally occurs through indirect liquefaction using an autothermal reactor. A recent review of hybrid and single feedstock energy processes can be found in [1]. A coal, biomass, and natural gas to liquids (CBGTL) thermochemical superstructure was recently developed [2-9] that incorporates various technologies, unit operating conditions, and stream interconnections. A process synthesis mathematical model was proposed to determine the process topology that can produce fuels at the lowest levelized cost [4, 6-8]. The hydrocarbon conversion route was primarily based on a combination of Fischer-Tropsch processes operating at multiple temperature levels, using either iron or cobalt catalysts.

This superstructure is expanded to include conversion from syngas to fuels via methanol intermediate, followed by the methanol-to-gasoline and methanol-to-olefins/distillate technologies to produce gasoline, diesel, and kerosene. The hydrocarbons can be upgraded to fuel quality in a standard upgrading section akin to the operation of petroleum refineries, or through catalytic conversion over ZSM-5 catalysts. Formulated as a mixed-integer nonlinear optimization problem (MINLP), the process synthesis problem is solved with simultaneous heat, power, and water integration.

Case studies are presented to investigate the trade-offs between the Fischer-Tropsch and methanol routes, along with detailed parametric analysis on the effect of plant capacities on the technology selection and optimal topologies. These examples elucidate the process topological differences and their effect of the overall system cost, the process material/energy balances, and the well-to-wheel greenhouse gas emissions.

[1] C.A. Floudas, J. A. Elia, R. C. Baliban (2012) Hybrid and Single Feedstock Energy Processes for Liquid Transportation Fuels: A Critical Review. Comp. Chem. Eng. 41:24-51.

[2] R. C. Baliban, J. A. Elia, C. A. Floudas (2010) Toward novel biomass, coal, and natural gas processes for satisfying current transportation fuel demands, 1: Process alternatives, gasification modeling, process simulation, and economic analysis. Ind. Eng. Chem. Res. 49:7343-7370.

[3] J. A. Elia, R. C. Baliban, C. A. Floudas (2010) Toward novel biomass, coal, and natural gas processes for satisfying current transportation fuel demands, 2: Simultaneous heat and power integration. Ind. Eng. Chem. Res. 49:7371-7388.

[4] R. C. Baliban, J. A. Elia, C. A. Floudas (2011) Optimization framework for the simultaneous process synthesis, heat and power integration of a thermochemical hybrid biomass, coal, and natural gas facility. Comp. Chem. Eng. 35:1647-1690.

[5] J. A. Elia, R. C. Baliban, X. Xiao, C. A. Floudas (2011) Optimal energy supply network determination and life cycle analysis for hybrid coal, biomass, and natural gas to liquid (CBGTL) plants using carbon-based hydrogen production. Comp. Chem. Eng. 35:1399-1430.

[6] R. C. Baliban, J. A. Elia, V. Weekman, C. A. Floudas (2012) Process synthesis of hybrid coal, biomass, and natural gas to liquids via Fischer-Tropsch synthesis, ZSM-5 catalytic conversion, methanol synthesis, methanol-to-gasoline, and methanol-to-olefins/distillate technologies. Submitted for publication.

[7] R. C. Baliban, J. A. Elia, C. A. Floudas (2012) Simultaneous process synthesis, heat, power, and water integration of thermochemical hybrid biomass, coal, and natural gas facilities. Comp. Chem. Eng. 37:297-327.

[8] R. C. Baliban, J. A. Elia, R. Misener, C. A. Floudas (2012) Global Optimization of a MINLP Process Synthesis Model for Thermochemical Based Conversion of Hybrid Coal, Biomass, and Natural Gas to Liquid Fuels. Comp. Chem. Eng. In press. DOI:10.1016/j.compchemeng.2012.03.008

[9] J. A. Elia, R. C. Baliban, C. A. Floudas (2012) Nationwide Supply Chain Analysis for Hybrid Feedstock Energy Processes with Significant CO_2 Emissions Reduction. AIChE Journal. Submitted for publication.

(445d) Ab-Initio and Computer-Aided Molecular Design for the Identification of Optimal Solvents for Reactions

Eirini Siougkrou, Chemical Engineering, Imperial College London, Centre for Process Systems Engineering, London, United Kingdom, Amparo Galindo, Department of Chemical Engineering, Centre for Process Systems Engineering, Imperial College London, London, United Kingdom and Claire S. Adjiman, Centre for Process Systems Engineering, Imperial College London, London, United Kingdom

A solvent is frequently used as a medium for liquid-phase reactions and its role is significant, especially in the control of reaction rates and temperature. The prediction of the effects of solvents on a given reaction and the rational choice of the most appropriate solvent for the reaction are complex challenges. Strübing et al.[1] have proposed an algorithm for the Computer-Aided Molecular Design (CAMD) of solvents for chemical reactions, where quantum mechanical (QM) calculations and a surrogate (approximate) model are used for the prediction of the reaction rate constant in various solvents. This approach has been successfully applied to the design of an improved solvent for a Menschutkin reaction, which was subsequently tested experimentally and shown to lead to a 40% increase in the measured rate constant, relative to the best initial solvent. Here we develop the approach further by removing the need for a surrogate model, which can be shown to introduce modeling error in the overall methodology. An optimisation CAMD formulation is used to predict the solvent that maximises the reaction rate. A design of several thousand potential molecules can be explored in this way. The reaction rate constants in the solvents tested during the course of the algorithm are obtained from QM calculations, using conventional transition state theory (CTST). The rate constant, according to CTST, is a function of the standard state free energy of solvation of the reactants and transition state. The free energy of solvation is calculated in this work from the SMD solvation model[2]. The proposed approach is compared to the CAMD algorithm of Strübing et al.[1] through application to a Menschutkin S_N2 reaction and the numerical performance and accuracy of both approaches is discussed.

References

[1] H. Strübing, P. G. Karamertzanis, E. N. Pistikopoulos, A. Galindo, C. S. Adjiman, Solvent design for a Menschutkin reaction by using CAMD and DFT calculations, Comp. Aided Chem. Eng., **2010**, 28, 1291-1296

[2] A. V. Marenich, C. J. Cramer, D. G. Truhlar, Universal solvation model based on solute electron density and on a continuum model of the solvent defined by the bulk dielectric constant and atomic surface tensions, J. Phys. Chem. B, **2009**, 113, 6378-6396

(445e) Dynamic Simulation and Load-Following Control of an IGCC Power Plant with CO2 Capture

Debangsu Bhattacharyya[1], Richard Turton[1] and Stephen E. Zitney[2], (1)Department of Chemical Engg., West Virginia University, Morgantown, WV, (2)AVESTAR Center, National Energy Technology Laboratory, Morgantown, WV

Load-following control of future integrated gasification combined cycle (IGCC) plants with pre-combustion CO_2 capture is expected to be far more challenging as electricity produced by renewable energy is connected to the grid and strict environmental limits become mandatory requirements. To study control performance during load following, a plant-wide dynamic simulation of a coal-fed IGCC plant with CO_2 capture has been developed. The slurry-fed gasifier is a single-stage, downward-fired, oxygen-blown, entrained-flow type with a radiant syngas cooler (RSC). The syngas from the outlet of the RSC goes to a scrubber followed by a two-stage sour shift process with inter-stage cooling. The acid gas removal (AGR) process is a dual-stage physical solvent-based process for selective removal of H_2S in the first stage and CO_2 in the second stage. Sulfur is recovered using a Claus unit with tail gas recycle to the AGR. The recovered CO_2 is compressed by a split-shaft multistage compressor and sent for sequestration after being treated in an absorber with triethylene glycol for dehydration. The clean syngas is sent to two advanced "F"-class gas turbines (GTs) partially integrated with an elevated-pressure air separation unit. A subcritical steam cycle is used for heat recovery steam generation. A treatment unit for the sour water strips off the acid gases for utilization in the Claus unit. The steady-state model developed in Aspen Plus® is converted to an Aspen Plus Dynamics® simulation and integrated with MATLAB® for control studies. The results from the plant-wide dynamic model are compared qualitatively with the data from a commercial plant having different configuration, operating condition, and feed quality than what has been considered in this work.

For load-following control, the GT-lead with gasifier-follow control strategy is considered. A modified proportional–integral–derivative (PID) control is considered for the syngas pressure control. For maintaining the desired CO_2 capture rate while load-following, a linear model predictive controller (LMPC) is implemented in MATLAB®. A combined process and disturbance model is identified by considering a number of model forms and choosing the final model based on an information-theoretic criterion. The performance of the LMPC is found to be superior to the conventional PID control for maintaining CO_2 capture rates in an IGCC power plant while load following.

(445f) Design of Transesterification Processes

Cory Silva[1], Leonard Fabiano[1] and **Warren D. Seider**[2], (1)Chemical and Biomolecular Engineering, University of Pennsylvania, Philadelphia, PA, (2)Chemical and Biomolecular Engineering, The University of Pennsylvania, Philadelphia, PA

DESIGN OF TRANSESTERIFICATION PROCESSES

by

Cory Silva, Leonard A. Fabiano, and Warren D. Seider

Department of Chemical and Biomolecular Engineering

University of Pennsylvania

Philadelphia, PA 19104-6393

Abstract

Biofuels, derived from organic oils, have received considerable attention during the last decade; however, most of the research effort has concentrated upon bench-scale experiments, with little or no attention given to the large-scale industrial applications of the various conversion methods. Despite the numerous techniques available,[1,2]current process research has focused mainly upon the use of aqueous alkaline catalysts[3], with some attention being given to non-catalytic, supercritical transesterification[4]. The narrow focus is likely due to the scarcity of kinetic data, which is almost nonexistent for non-aqueous catalysts.

In this paper, an algae-oil transesterification process is presented, using a proprietary solid-acid catalyst. Kinetic parameters are regressed from bench-scale experimental data for oil harvested from *Nanochloropsis Salina* algae. The process model is simulated using ASPEN PLUS, and is then heat integrated to reduce the consumption of utilities. The equipment is sized and costed using ASPEN's Process Economic Evaluator to establish an overall project cost. Finally, our model and costs are compared with previous studies for algae-derived oils[5] and vegetable oils[3]to assess the viability of a process using heterogeneous catalysis.

References: 1. Vyas A. P., Verma J.L., Subrahmanyam N., 2010. A review on FAME production processes, *Fuel*, **89**, 1–9. 2. Demirbas A., 2008. Comparison of transesterification methods for production of biodiesel from vegetable oils and fats, *Energy Conversion and Management*, **49**, 125–130. 3. Chang A., Liu Y.A., 2010. Integrated process modeling and product design of biodiesel manufacturing, *Ind. Eng. Chem. Res*., **49**, 1197-1213. 4. van Kasteren J.M.N., Nisworo A.P., 2007. A process model to estimate the cost of industrial scale biodiesel production from waste cooking oil by supercritical transesterification, *Conservation and*

Recycling, **50**, 442–458. 5. Pokoo-Aikins G., Nadim A., El-Halwagi M.M., Mahalec V., 2010. Design and analysis of biodiesel production from algae grown through carbon sequestration, *Clean Techn. Environ.*

Policy, **12**, 239-254.

(445g) Comparison of Extractive Distillation and Pressure-Swing Distillation for Methanol-Chloroform System

Eda Hosgor and **Devrim B. Kaymak**, Department of Chemical Engineering, Istanbul Technical University, Istanbul, Turkey

Two of the most common methods for separating a binary homogeneous azeotrope are extractive distillation and pressure-swing distilllation. The first method is effective if a suitable entrainer can be found, while the second one is effective if the composition of the azeotrope changes significantly with pressure. In this study, the steady-state designs of these methods will be developed for a methanol-chloroform binary system which is an effective solvent for extraction of bioactive substances from biological sources. For this binary system extractive distillation is feasible using water as the entrainer. Also the pressure-swing distillation is feasible since the minimum-boiling azeotrope at 1 atm contains 63,32 mol % chloroform at 327 K and at 10 atm the azeotropic composition is 40,83 mol % chloroform at 404 K. For all the process simulations the feed mixture contains 50 mol % chloroform 50 mol % methanol with a flowrate of 100 kmol/h and the purities of the two products are set at 99.5 mol %. Wilson model is used as thermodynamic model. For the extractive distillation process, the design variables to be optimized are; entrainer flowrate, extractive distillation total number of stages, fresh feed and entrainer feed locations, entrainer recovery column total number of stages, feed location of entrainer recovery column. For the pressure swing distillation process, the design variables to be optimized are; total number of stages and feed locations for both columns and recycle flowrate. The aim of this study is the comparison of steady-state designs of these methods by investigating the effects of design variables and total annual costs.

(466a) Application of Quality by Design Principles for An Existing Product for the Control of the Dimer Impurity in the Manufacturing Process

Gautam Samanta[1], Praveen K Sharma[2], S Srividhya[2], Kamlesh Pai Fondekar[2], K R Janardan Sarma[2], Saurabh Gupta[2], Grb Ravish[2] and Y Raghuram[3], (1)Dr. Reddy's Laboratories Ltd., Hyderabad, India, (2)Dr. Reddy's Lab, Hyderabad, India, (3)DR. Reddy's Lab, Hyderabad, India

Application of Quality by Design principles for an existing product for the Control of the Dimer Impurity in the Manufacturing Process

Gautam Samanta, Praveen K Sharma, S Srividya, Kamlesh Pai Fondekar, K R Janardan Sarma, Saurabh Gupta, GRB Ravish, and Y Raghuram

Dr. Reddy's Laboratories

Integrated Product Development Organization

Bachupally, Hyderabad

India

The philosophy of quality by design is all about the patients' need. The designing, developing, and manufacturing of the product must meet those needs. Most of the innovator and few generic companies have adopted the concept of QbD in their product development. Dr. Reddy's Laboratories has started implementing the QbD principles for all new APIs development. The QbD implementation for new products is well documented but not for existing products. The key drivers for the implementation of QbD for the existing products are to reduce variability in the product quality, improve yield, reduce cycle time, solve manufacturing issues, reduce cost of quality, and introduction of real time release testing for manufacturing process.

The objective of this work is to provide a perspective of QbD application for existing API manufacturing where a stepwise reverse QbD approach has been applied. The unique features of this work comprise methodology and implementation strategy in defining design space and thereafter executing the control strategy for consistent production of the API.

Irbesartan is an angiotensin II receptor antagonist used mainly for the treatment of hypertension. Dr. Reddy's Laboratories developed the generic version and filed DMF in March 2009. In stage 1, IRB-1 is formed by reaction between 2 n butyl-4-spiro cyclopentane-2-imidazoline-5-one and 4-bromo methyl-2'-cyanobiphenyl at -5 to -10 °C in the alkaline medium (Figure 1). The reaction time is 2-3 h. In stage 2, IRB-1 is converted to tetrazole by the reaction with NaN_3 at elevated temperature and the final product Irbesartan (IRB-2) is obtained after crystallization.

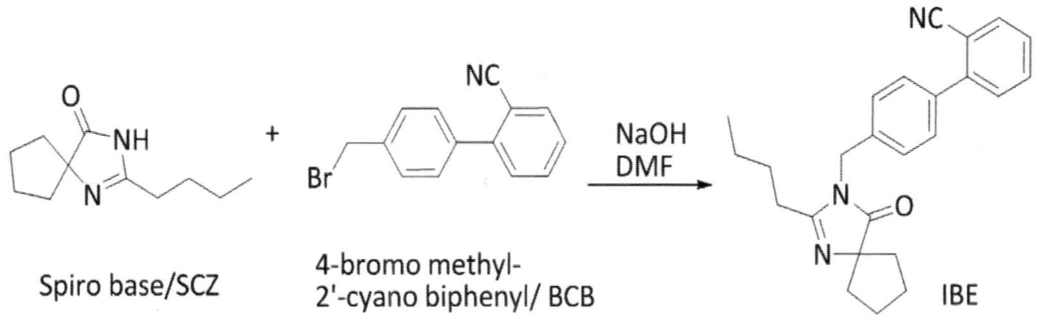

Figure 1: Stage 1 for the synthesis of Irbesartan and the structure of the Dimer impurity (CQA)

In stage 1, there is a significant variation (upto 5%) in the formation of a dimer impurity, which is difficult to remove in the subsequent stages. If the dimer content is more than 1%, significant yield loss occurs for its removal. So the cost of quality is high. The dimer impurity is the critical quality attributes (CQA) in this stage. The QbD principles have been applied to address the issue.

The following stepwise QbD approaches has been adopted:

1. Historical data and analysis of data to identify the problem

There is a significant variation in the formation of the dimer in the process. So some variables are playing critical role. To understand the variability in the dimer quantity experiments were conducted at pre-determined time under the same experimental conditions.Figure 2 shows the experimental results, which indicate that the formation of the dimer depends on the reaction time. Even within the reaction time 2-3 h, the dimer quantity is varying significantly and the variation is more after 3 h which indicates that reaction time is very critical. The best fit with the data is the third-order polynomial equation (R^2= 0.994).

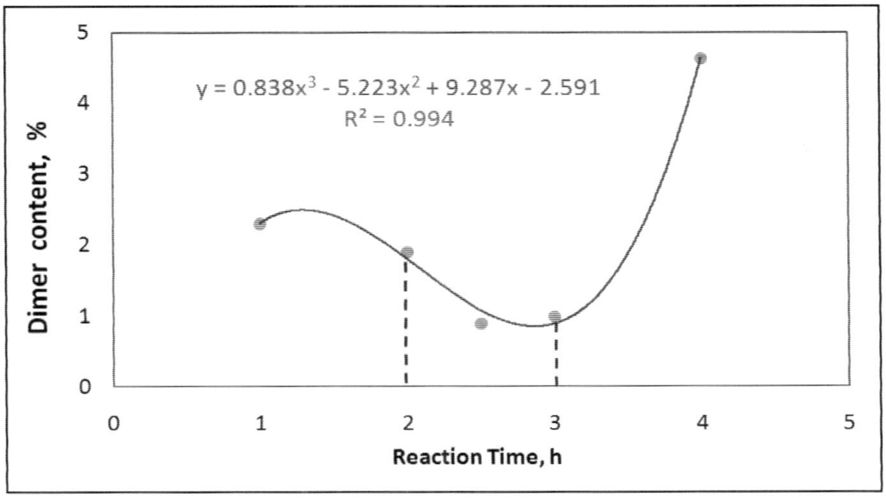

Figure 2: Best fitting of the experimental data

2. Quality risk assessment for material attributes and process parameters

To identify the critical variables the cause-effect diagram (Figure 3) has been used. All material attributes and process parameters have been listed in this diagram. Based on the chemistry knowledge as well as practical experience it has been concluded that material attributes are not responsible for the dimer. So only the process parameters have been considered for risk analysis.

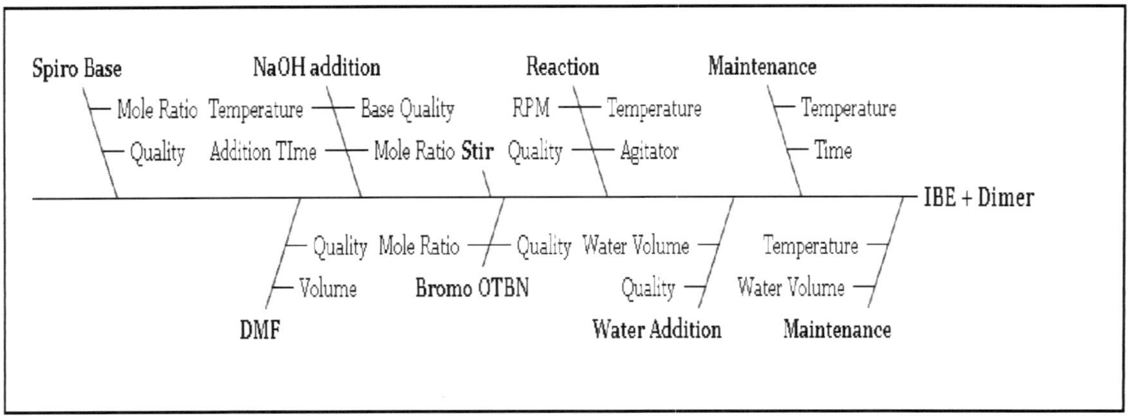

Figure 3: Cause-effect or fishbone diagram for MA and PP to understand the effect on CQAs

In the case of PPs, the first phase is to identify those potential CPPs that may affect the CQAs of the API (*i.e.*, the severity). The second phase of the assessment is to estimate the probability of a parameter affecting the API (*i.e.*, the risk). The parameters with a high probability of influencing the API with enough severity to produce unacceptable API are critical parameters. The overall risk is the combination of severity and probability and risk assessment is the impact of the process on the API. Table 1 shows the risk assessment of the process parameters.

Table 1: Risk assessment for the process parameters

Trigger event	Severity	Probability	Overall Risk	Risk assessment
DMF volume	Low	Low	Low	Low risk to Dimer
Water volume	Low	Low	Low	Low risk to Dimer
H_2O addition temp.	Low	Medium	Medium Low	May exceed range; no impact on Dimer
Reaction temp.	Low	Medium	Medium Low	Parameter is critical
RPM	High	Medium	Medium High	May impact on Dimer; Parameter is critical
NaOH	High	Medium	Medium High	May impact on Dimer; Parameter is critical
Reaction time	High	High	High	High impact on Dimer; Parameter is critical

Based on the preliminary risk analysis of the process parameters, RPM, NaOH quantity, and reaction time have been identified as critical process parameters. Though the reaction temp is critical, it has not been considered for DoE study because the reaction is not completing above -5 °C and the reaction mass is freezing below -18 °C. Due to a narrow window of operation it was decided to fix it at -7.5 ± 2.5 °C. To establish the relation between the critical process parameters and CQAs the design of experiments (DoE) has been applied.

3. DoE, understanding of process, and design space

Initially, two-level factorial design was selected and twelve experiments (2^3 + 4 center points) were conducted in the lab considering three variables and their respective ranges as mentioned in Table 2. The normal plot for the dimer

(Figure 4) reveals that all the three variables are making significant positive impact on the dimer formation (NaOH 36%, RPM 21%, and Reaction time 13%). It is interesting to note that the three-variable interaction (ABC) has negative impact (15%) on the dimer formation. Due to significant curvature a central composite design was selected and remaining experiments were conducted.

Table 2: Critical process parameters and their ranges

Variables	Low	High	Rational for Selection
NaOH	1.2 eq.	1.7 eq	Kinetic Effect: Higher amount of base could increase the rate of formation of dimer.
Reaction Time	150 min	200 min	Kinetic Effect: Longer reaction time could increase the formation of dimer
RPM	300	600	In heterogeneous system, RPM will influence the heat/mass transfer and distribution of solid in liquid medium.

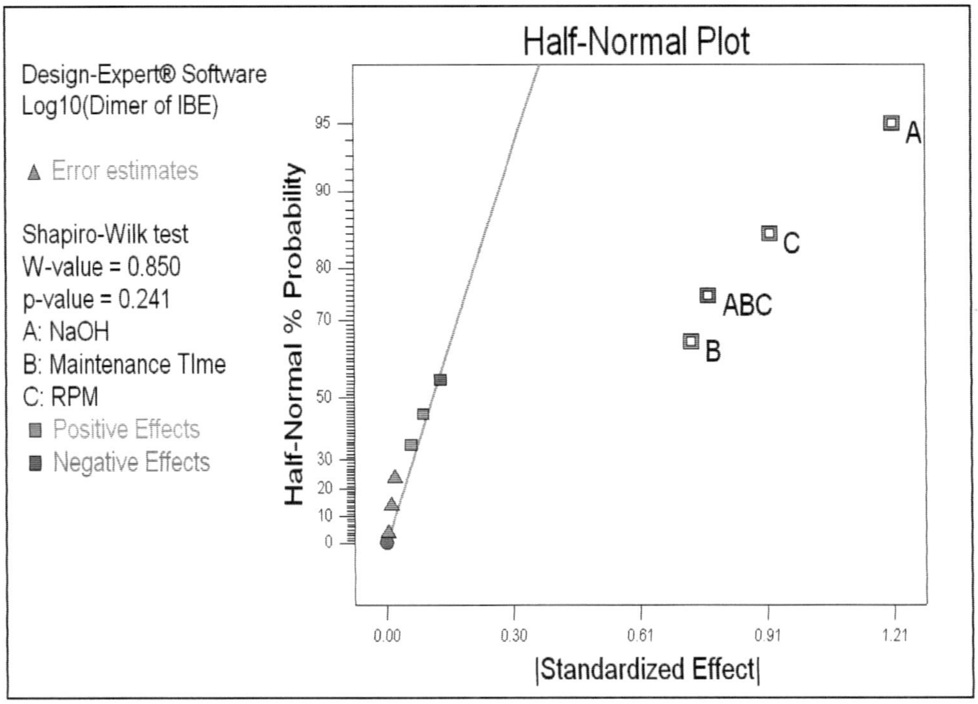

Figure 4: Half-normal plot for two-level factorial design

The data from CCD are fitted to a cubic model (R^2 = 0.9999, Adj R^2 = 0.9997, Pred R^2 = 0.9496, and PRESS = 2.76). Analysis of variance testing (Table 3) confirms the significance of three-factor interaction (ABC), as well as of other terms A^2B, AB^2, A^2C, A^2, B^2, C^2 and two-factor interactions AB, AC, and BC. The three-dimensional surface plot, illustrated in Figure 5 is generated according to the cubical model and demonstrates the dependence of dimer on the amount of NaOH and RPM at optimal level of reaction time (2.5 h). To see the influence of three-factor interaction

(ABC), the model term (ABC) has been excluded and the analysis of variance indicates that in the absence of that term almost all the terms become insignificant (Table 4) and the lack of fit becomes significant. This indicates that the three-factor interaction term is playing very critical role in the formation of dimer.

Figure 6 is an overlay plot of yield, dimer, and other QAs, setting the foundation for continued improvement of the process. The yellow region is design space where the desired QA levels will be obtained. Laboratory validation experiments were performed under the experimental conditions (NaOH 1.3 eq, Reaction time 200 ± 10 min, RPM 300 ± 20) chosen from the design space and the dimer (0.29 ± 0.06) and other quality attributes were within the specifications.

Table 3: ANOVA table and associated statistics for response surface cubic model for dimer

ANOVA for Response Surface Reduced Cubic Model

Analysis of variance table [Partial sum of squares - Type III]

Source	Sum of Squares	df	Mean Square	F Value	p-value Prob > F	
Model	54.68	13	4.21	5578.23	< 0.0001	significant
A-NaOH	*1.33*	*1*	*1.33*	*1761.66*	*< 0.0001*	
B-Maintenance	*0.95*	*1*	*0.95*	*1262.71*	*< 0.0001*	
C-RPM	*0.31*	*1*	*0.31*	*413.81*	*< 0.0001*	
AB	*0.065*	*1*	*0.065*	*85.93*	*< 0.0001*	
AC	*4.74*	*1*	*4.74*	*6289.96*	*< 0.0001*	
BC	*0.22*	*1*	*0.22*	*288.82*	*< 0.0001*	
A^2	*1.82*	*1*	*1.82*	*2408.79*	*< 0.0001*	
B^2	*1.30*	*1*	*1.30*	*1724.81*	*< 0.0001*	
C^2	*0.38*	*1*	*0.38*	*506.63*	*< 0.0001*	
ABC	*3.67*	*1*	*3.67*	*4869.51*	*< 0.0001*	
A^2B	*0.11*	*1*	*0.11*	*151.83*	*< 0.0001*	
A^2C	*1.26*	*1*	*1.26*	*1671.22*	*< 0.0001*	
AB^2	*0.55*	*1*	*0.55*	*732.34*	*< 0.0001*	
Residual	4.525E-003	6	7.541E-004			
Lack of Fit	*2.241E-003*	*1*	*2.241E-003*	*4.91*	*0.0776*	*not significant*
Pure Error	*2.283E-003*	5	*4.567E-004*			
Cor Total	54.69	19				

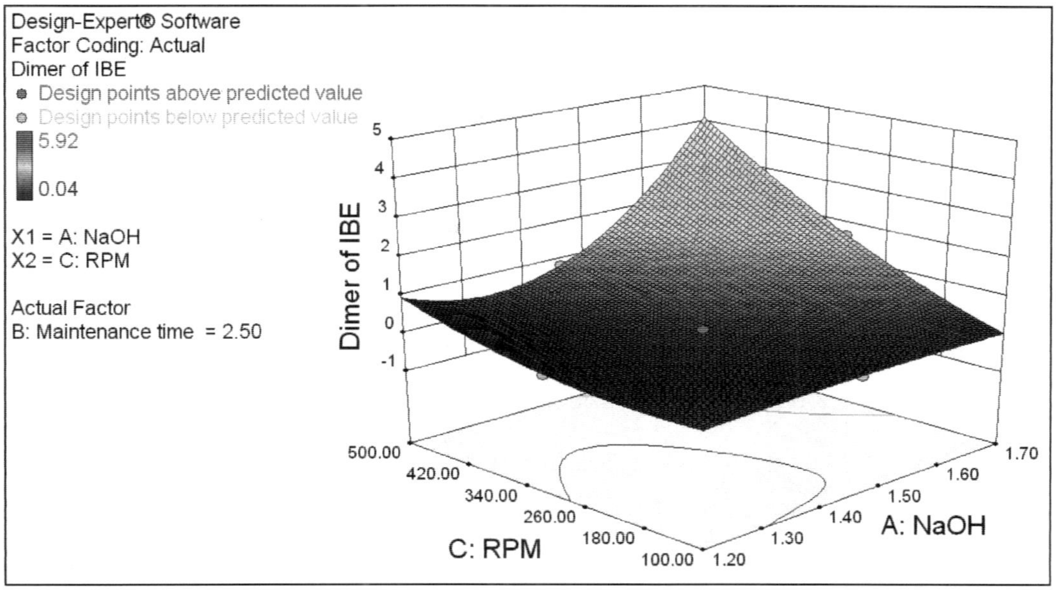

Figure 5: Influence of NaOH and RPM on the formation of dimer impurity at reaction time 2.5 h

Table 4: ANOVA table and associated statistics for response surface cubic model for dimer excluding ABC interaction term

ANOVA for Response Surface Reduced Cubic Model

Analysis of variance table [Partial sum of squares - Type III]

Source	Sum of Squares	df	Mean Square	F Value	p-value Prob > F	
Model	51.01	12	4.25	8.09	0.0052	significant
A-NaOH	*1.33*	*1*	*1.33*	*2.53*	*0.1558*	
B-Maintenance	*0.95*	*1*	*0.95*	*1.81*	*0.2201*	
C-RPM	*0.31*	*1*	*0.31*	*0.59*	*0.4660*	
AB	*0.065*	*1*	*0.065*	*0.12*	*0.7357*	
AC	*4.74*	*1*	*4.74*	*9.03*	*0.0198*	
BC	*0.22*	*1*	*0.22*	*0.41*	*0.5401*	
A^2	*1.82*	*1*	*1.82*	*3.46*	*0.1053*	
B^2	*1.30*	*1*	*1.30*	*2.48*	*0.1596*	
C^2	*0.38*	*1*	*0.38*	*0.73*	*0.4219*	
A^2B	*0.11*	*1*	*0.11*	*0.22*	*0.6548*	
A^2C	*1.26*	*1*	*1.26*	*2.40*	*0.1653*	
AB^2	*0.55*	*1*	*0.55*	*1.05*	*0.3393*	
Residual	3.68	7	0.53			
Lack of Fit	*3.67*	*2*	*1.84*	*4022.95*	*< 0.0001*	*significant*
Pure Error	*2.283E-003*	*5*	*4.567E-004*			
Cor Total	54.69	19				

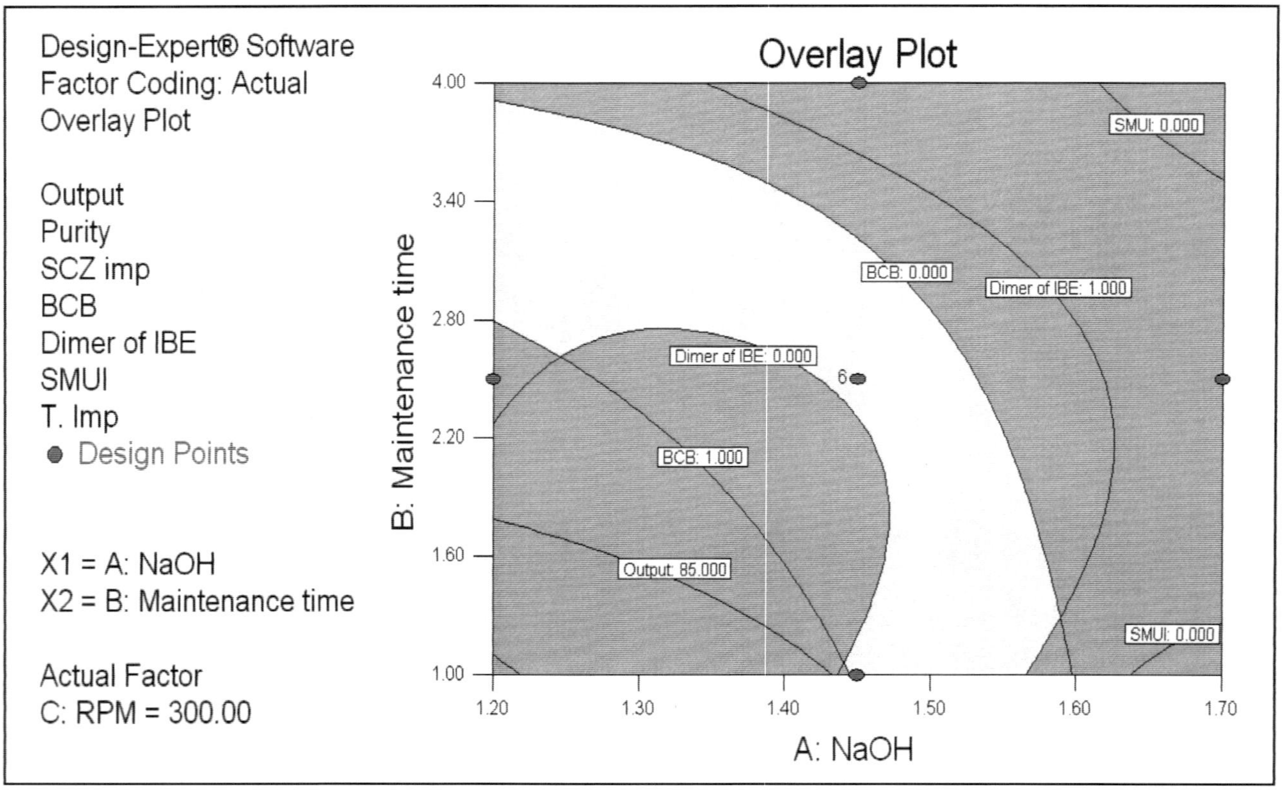

Figure 6: Overlay plot of Yield, starting materials, purity, dimer impurity, and single maximum unknown impurity for stage 1, which provides a process operating window

4. Control strategy

Does the output from the laboratory-scale multivariate design get reflected during scale-up? It is important to do the appropriate scale-up for the scale dependent variables before manufacturing. Based on the process understanding and criticality a control strategy has been developed to consistently ensure product quality. To control the formation of the dimer in the plant the scale dependent variable, RPM was scaled-up based on tip-speed and control on other variables was achieved through automation. In-process analysis was not available to monitor the formation of the dimer at that time. The consecutive six batches data in the plant indicates that the dimer is under control within the specification (<1%).

The QbD principles provide a structured approach for gaining process knowledge and developing robust manufacturing process. The application of the QbD principles for the existing products is very limited. This has been applied successfully to reduce the variability of the dimer impurity in the process and reduces the cost of quality. On the basis of the process understanding, it has been demonstrated that the control of the dimer impurity can be achieved by controlling the critical process variables.

(466b) Investigation of the Impact of Mode of Operation (Batch vs Flow) On Design Space Map

Shankar Vaidyaraman, Sze Wing Wong and Shekhar K. Viswanath, Chemical Product R&D, Eli Lilly and Company, Indianapolis, IN

This talk will discuss how the mode of operation (batch vs flow) impacts the design space map for impurities thus impacting the robustness of the process. A specific case study involving a reaction model will be used for this investigation that builds upon previous work on failure mode analysis in continuous processes (Wong *etal* 2011). There are three impurities of interest formed in this reaction. A reaction kinetic model developed from batch experiments is used to design for both batch and flow processes. The key questions that this investigation addresses is:

- How much does the mode of operation impact the size of the design space?
- Is the link between the uncertainties due to routine variability and model uncertainty and the output impurity levels same or different depending on the mode of operations?
- Can certain generalizations be drawn based on the results of this case study?

[1] Sze-Wing Wong, Prashant Kokitkar, K. Derek Berglund, Martin D Johnson, and Shekhar K. Viswanath, Failure Mode Analysis of Continuous Process with Process Models, AIChE Annual Meeting, MN 2011

(466c) A QbD Approach to the Process Development of Tofacitinib

Bob Dugger, Chemical Research and Development, Pfizer Inc, Groton, CT

Tofacitinib citrate (CP-690,550-10) is currently being evaluated for the treatment of rheumatoid arthritis, psoriasis and other diseases. An NDA has recently been submitted to the FDA for the treatment of rheumatoid arthritis. The Pfizer Tofacitinib API Team has worked to improve the efficiency of the process and ensure that the quality of the drug substance will meet the standards of the regulatory agencies. Quality by Design is important not only from a regulatory perspective, but also from a process development perspective since it enforces a much more thorough degree of process understanding. DOE is often the focus of QbD but there are often other techniques (experimental and modeling) that can be used in the development of a process. This talk will provide examples of how we use these various techniques to ensure we have a robust process that can be run at a wide range of scales and consistently meet our quality standards.

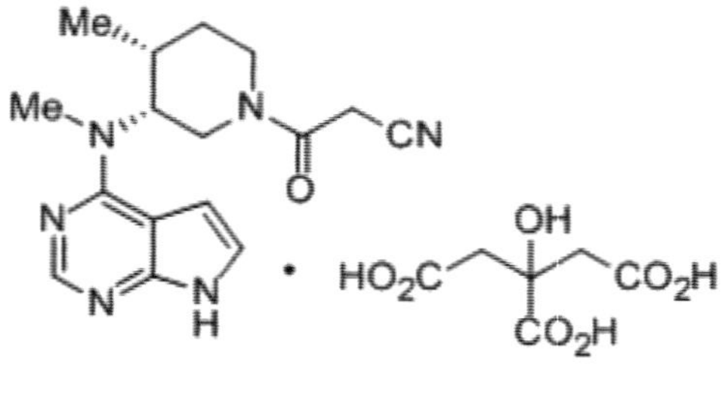

Tofacitinib citrate

(466d) Design of a Pharmaceutical Manufacturing Process Using QbD Principles: The Challenge of Technical Transfer to Manufacturing and Lifecycle Development

Conor Collins and Steve Ball, GlaxoSmithKline, Cork, Ireland

The route of manufacture for product X is a convergent, multi-stage chemical synthesis that has been developed using QbD principles.

A description of the approach to chemical process development utilising QbD methodologies, to determine and control quality critical process parameters is provided. A workflow from process assessment, through process understanding to development of a control strategy, providing a platform for technical transfer to commercial manufacturing is described.

The process involved in the development of a commercial manufacturing strategy to utilise the thorough process understanding, in order to ensure the provision of a robust commercial manufacture is also illustrated.

Rising to the challenge of providing a robust chemical synthesis, adequately described in the registered manufacturing description in combination with supporting compliance during the product's commercial lifecycle is essential.

(466e) Strategy and Process for Effective Scale-up On the Removal of Pd During the Synthesis of A Drug Substance

Alexander Chu-Kung, Global Pharmaceutical Research and Development, Abbott Laboratories, North Chicago, IN and Hsien-Hsin Tung, Global Pharmaceutical Research & Development, Abbott Laboratories, North Chicago, IL

Metal catalyst, for example Pd or its derivatives, are commonly used during the synthesis of small molecule drug substance. Due to the toxic nature of these catalysts, their levels are highly controlled at the final drug substance stage. Generally, rejection of metal catalyst is not easy and requires multiple steps of purifications. However, from time to time the approach of metal removal could be highly empirical with a minimum understanding on the fundamentals. Given than, it would be very desirable to address the issue through a more fundamental approach to ensure the efficiency and consistency of catalyst removal upon scale-up.

In our presentation, we share our experience during the development of drug substance for the removal of Pd catalyst. A dual approach, incorporating both the solubilization method and adsorption method are employed and fully integrated into the process. For solubilization, the critical control points are identified to be mixing and cake washing. For adsorption, the most cost-effective adsorbents are shown to be carbon and filtrol. Pd rejection consistency and efficiency are fully demonstrated at the pilot plant scale after implementing the proper procedures with the findings.

(466f) Implementation of Reaction Modeling and Process Analytical Technology in Design Space Development

Daniel Hallow, Michael Fenster, Antonio Ramirez, Robert Wethman and Nathan Domagalski, Bristol-Myers Squibb Company, New Brunswick, NJ

A key aspect of Quality by Design (QbD) in pharmaceutical manufacturing is the development of a design space for multivariate processes. In this case study, we explored the application of mechanistic modeling and process analytical technology (PAT) to define the design space of a reaction that generates the pharmaceutical intermediate of a drug substance in development.

The reaction involves a nucleophilic aromatic substitution that affords the desired material along with two critical impurities. The main parameters impacting the formation of the impurities are the excess base and the temperature of the reaction. A reaction model based on mechanistic studies was built using data captured by HPLC and then used to conduct simulations that generated the response surfaces corresponding to the conversion of the starting materials and the formation of the critical impurities. In addition, the reaction progress was analyzed with Raman spectroscopy and a chemometric model was built to allow real-time monitoring of the starting material conversion. Combining the knowledge of the reaction response across multiple parameters with Raman spectroscopy for real-time monitoring of the reaction progress allows for relaxation of processing constraints and resulting in the expansion of the design space.

In summary, this work shows the synergistic benefits of combining mechanistic modeling and PAT in a QbD approach to enhance the design space and robustness of a process.

Application of Quality by Design in API Process Development

Umetrics – An MKS company
Erik Johansson, Senior Application Specialist
Nirav Shah, Senor Application Specialist

Design of Experiments (DOE) is the most effective method to achieve product and process efficiency and optimization. Design of Experiment (DOE) studies can help develop *process development knowledge* by revealing relationships, including multifactorial interactions, between the variable inputs (e.g., component characteristics or processing parameters) and the resulting outputs (e.g., in-process material, intermediates, or the final product).

QbD should by definition involve DoE to identify critical interactions and noise. There are many steps in establishing a Design Space for QbD. This presentation will focus on the actual calculation and determination of a Design Space with many factors. The ICH Q8 document gives an illustrative example of how to handle this for two factors and broadly describes how additional factors should be treated. In an example based presentation we will demonstrate how the design space can be calculated for 5 factors and also illustrate the fundamental difference between a design space approach and a process set-point with a maximum process window. The process set-point with 5 factors gives a high and low value for each factor corresponding with ICH Q8. The Monte Carlo derived design space approach is more in accordance with ICH Q11 where the actual design space is given by an equation. The advantage with this approach is that it gives the possibility for control-meaning that the best value for one factor is changed on the basis of the set point of another factor. In the graph below we can see that when the Substrate Concentration is 1.7, the best values for NaOH and CaCl2 are 1.6 and 1.4. Similarly at Substrate Concentration 1eq the corresponding values are 1.55 and 1.2.

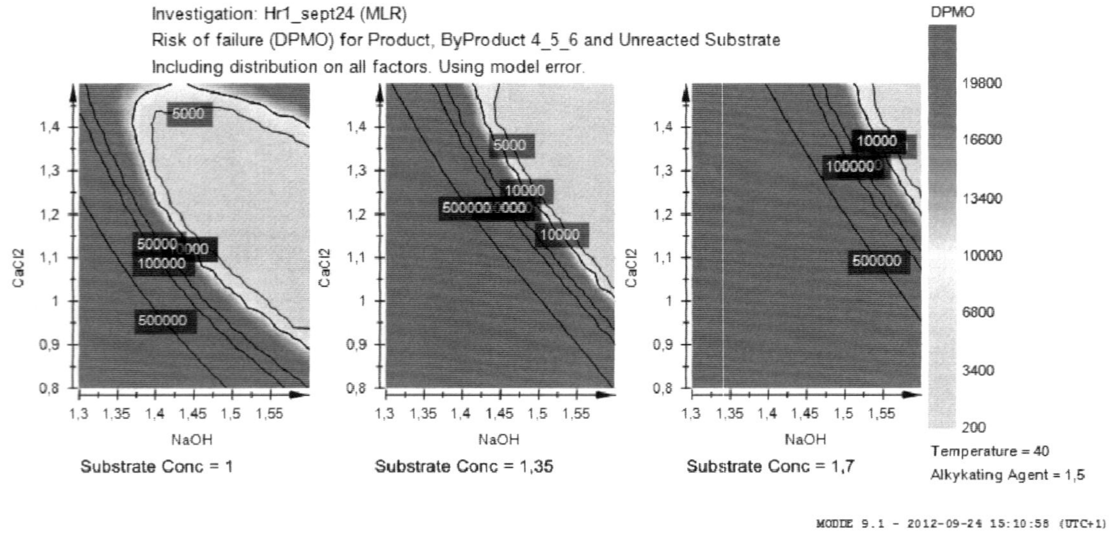

(533a) Drug Product Model Design, Development, Qualification, and Maintenance in a QbD Framework

Hayden Thomas, Formulation Development, Vertex, Cambridge, MA

Model development and implementation is a core principle to employing a QbD approach in drug product design and process scale-up. The advantages in taking this approach are realized through the ease of visualization or mathematical expression of data sets for greater process understanding, which leads to clear decision making and results in superior product quality. There are several model types that can be developed and used at every stage in the QbD process. A favored approach is to begin with the end in mind by outlining up front a clear plan toward model design and development for the drug product process, considering how each model will be used, limitations, assumptions, qualification, and maintenance. The experimental approach to the development of each model should consider the type of model (mechanistic or empirical), selected variables, scale dependency and relationship to the overall drug product process. Once a model has been developed, steps are needed toward implementation through qualification of the model and evaluating uncertainty in predictions. Finally, for models selected for use in commercial manufacture, a maintenance plan is needed to support product life cycle, allowing model updates as needed. In this presentation, representative examples will be provided using these principles that have been used to support drug product filings.

(533b) Prediction of Post-Relaxation Radial Tensile Strength of Pharmaceutical Tablets Using At-Line Collected near Infrared Data

Sameer Talwar[1,2], Benoit Igne[2], Zhenqi Shi[2], Carl A. Anderson[2] and James K. Drennen III[2], (1)Graduate School of Pharmaceutical Sciences, Duquesne University, Pittsburgh, PA, (2)School of Pharmacy, Duquesne University, Pittsburgh, PA

The prediction of relaxed radial tensile strength (RTS) of pharmaceutical tablets in real-time using near infrared (NIR) spectroscopic data collected at-line was investigated. Radial tensile strength is a critical quality attribute (CQA) of tablets which may correlate to drug product performance *in vivo*. However; post-compaction, the tablets undergo axial and radial relaxation, which influences dosage form characteristics such as disintegration time, dissolution time, and mechanical strength. It is the RTS of completely relaxed tablets that is of interest, as it is a completely relaxed tablet that will be administered to a patient. The existing work relating NIR spectra to RTS relies on the use of data collected from relaxed tablets (off-line NIR data) to predict the relaxed (off-line) RTS. However, the NIR analysis of tablets in a manufacturing setting, as desired in the current Quality by Design philosophy and accomplished in a modern real-time release environment, necessitates that tablet quality attributes are determined immediately following compression and prior to complete relaxation.

A six component system was compressed with variability in excipient ratio, disintegrant level, and compression force (total of sixteen design points). Eighteen tablets were collected per design point and scanned at-line using a commercial NIR spectrometer immediately after compression. Tablets were left to relax for two weeks. Their physical dimensions were then measured and they were crushed to calculate RTS values. A multiple linear regression (MLR) was used to predict RTS of relaxed tablets from at-line spectra. In a first step, partial least squares models were developed to predict variables with a significant relationship to RTS (tablet density, tablet volume). Outputs from the PLS models were then used along with spectral slope and formulation parameters as input in an MLR to predict RTS. Radial tensile strength predictions on an independent test set had an error of 0.25 MPa with a R^2 of 0.8. The MLR model had calibration and cross validation error between 0.2 – 0.3 MPa. Updating spectral information with physical features and chemical information through MLR allowed for predicting RTS of relaxed tablets using at-line information while compensating for the effect of tablet relaxation over time. This approach has the potential to be implemented on-line since all the information needed by the MLR model can be calculated from the same near infrared spectrum. Automating spectral collection will allow the real-time prediction of relaxed RTS at the tablet press level.

(533c) Quantitative Analysis of Crystalline Materials in Binary Compacts Using Time Domain Spectra of THz Spectroscopy and Density Dependence

Md. Anik Alam, James K. Drennen III and Carl A. Anderson, Mylan School of Pharmacy, Duquesne University, Pittsburgh, PA

Polymorph and pseudo-polymorph systems play a major role in the pharmaceutical industry because of their potential to impact the performance of finished dosage forms. Quantitative analysis of polymorph and pseudo-polymorph is becoming increasingly important for solid oral dosage forms and transdermal products. A rapid, non-destructive analytical tool is required to reduce the time and cost for quantitative analysis. Terahertz Time Domain Spectra has emerged as a potential analytical tool for rapid and non-destructive analysis of these systems. Terahertz Time Domain Spectra provides high specificity for polymorphic study due to the ability to probe inter-molecular and phonon lattice vibration directly into the crystal structure.

The feasibility of Terahertz Time Domain Spectra for quantitative analysis of anhydrous theophylline and the monohydrate in compacts was studied. This study included the analysis of density dependence in quantification accuracy. The effect of density on the Terahertz Time Domain Spectra was demonstrated. An improvement in model performance was demonstrated when density was undertaken into consideration.

(533d) Bulk and Micro-Scale Characterization of Continuous Powder Blending Integrated with an Multi-Point in-Line NIR As a PAT Tool

Juan G. Osorio, William E. Engisch Jr. and Fernando J. Muzzio, Chemical and Biochemical Engineering, Rutgers, The State University of New Jersey, Piscataway, NJ

Continuous manufacturing of solid pharmaceutical products has been an area of much research in the past few years. Most efforts have focused in characterizing the bulk properties and efficiency of continuous powder mixers as a function of operational and design parameters. The present study summarizes the bulk characterization of a new Glatt continuous powder (GCG-70) blender using experimental techniques previously reported [1,2]. While bulk characterization of a resulting blend is crucial to describe the mixing efficiency of a continuous mixer, it is not enough to determine the final quality of a blend, which can then be related to the performance in succeeding processes. A new approach to study final blends at the micro-scale using hyperspectral imaging to further understand phenomena such as API agglomeration and de-agglomeration in continuous mixing is reported here. Further along, comparison results from the integration of an in-line NIR as a PAT tool to monitor the concentration and variability of key components in final blends are described. Both approaches presented here were useful to further understand the continuous mixing process as a function of the design, operational and material properties parameters.

References:

[1] P.M. Portillo, M.G. Ierapetritou, F.J. Muzzio, Characterization of continuous convective powder mixing processes, Powder Technology 182 (2008) 368-378.

[2] A.U. Vanarase, F.J. Muzzio, Effect of operating conditions and design parameters in a continuous powder mixer, Powder Technology 208 (2011) 26-36

(533e) Effect of Lubricant Mixing Time and Concentration, Compaction Speed and Compression Force On the Dissolution Profile of a Control Release Formulation

Sarang Oka[1], Pallavi Pawar[2], Fani Boukouvala[2] and Fernando J. Muzzio[1], (1)Department of Chemical and Biochemical Engineering, Rutgers, The State University of New Jersey, Piscataway, NJ, (2)Chemical and Biochemical Engineering, Rutgers University, Piscataway, NJ

Effect of lubricant mixing time and concentration, compaction speed and compression force on the dissolution profile of a control release formulation

Sarang Oka, Pallavi Pawar, Fani Boukouvala, Fernando Muzzio

Rutgers University

Segregation or de-mixing of granular matter has been studied in great depth and in the pharmaceutical industry continues to be a routine nuisance. In the current project, significant variability was observed in the dissolution profile between tablets made from a multiparticulate formulation. The formulation in discussion is a typical formulation consisting of an API in the form of granulated pellets, filler, Magnesium Stearate and a rapid disintegrant. The particle size of the pellets is 600 μm. Tablet dissolution is performed in a standard USP dissolution apparatus-2. The dissolution profile demands no more than 20% of the drug to be dissolved in 60 minutes, no more than 35% in 120 minutes, between 20-45% in 480 minutes and no less than 80% in 1440 minutes.

The standard deviation in the value of the amount of drug dissolved in one hour was found to be unacceptably high by regulatory standards. The observation was attributed to segregation since the final formulation exhibits a significant tendency to demix.

In order to enable a QbD approach for solving this problem, preliminary studies were conducted to examine the effect of several variables on the dissolution profile of the tablets. It was hypothesized that the amount of lubricant and the amount of shear in lubricated mixture could have a significant impact on the observed response. Compression speed and compression force were also considered to be potentially important.

The work focused on conducting a preliminary set of designed experiments to test the relative importance of multiple variables on a lab scale, perform statistical analysis to determine significant parameters affecting dissolution and consequently develop a more concise Design of Experiment to determine the design space at the manufacturing scale. The original formulation contains 0.2% MgSt. The study was a full factorial DOE with 0.2%, 0.4% and 0.6% MgSt mixed for 50, 100 and 150 revolutions that result in 10 blends (the 10th blend employs 0.2% MgSt mixed for 300 revs). Mixing was performed in a conical twin-axial mixer at 30 rpm. The blend was sampled to test mixing homogeneity. Mixing homogeneity was quantified by performing offline NIR measurements on extracted samples. The blend is finally compacted in a tablet simulator at 3 speeds and 3 punch depths (compaction force) resulting in 9 compaction conditions per batch leading to 90 tested conditions in all.

Analysis of variance tests are performed on the ensuing data set. Statistical significance of compaction force, compaction speed, concentration of lubricant and mixing time of lubricant on percent drug dissolved in 60 minutes is tested. The compaction speed, lubricant concentration and lubricant mixing time are all found to be statistically significant parameters which affect the amount of drug dissolved in 60 minutes. Coupled interactions of the above parameters are found to be significant while three way interactions were found to be insignificant.

The dominant effects were the compression speed and the amount of lubricant (Magnesium Stearate in this case) in the formulation., leading to a large interactive effect with the compression force. At low levels of Magnesium stearate, an increase in compression force increased the dissolution rate, perhaps due to increased pellet breakage. On the other hand, at higher levels of Magnesium stearate, the rate of dissolution decreased with an increase in force, due to increased tablet consolidation.

QbD Case Studies: An Integrated Multivariate Approach to Drug Product and Process Development

Presenting Author: Goldi Kaul (Ph.D.)

(Authors: Jun Huang, Ramarao Chatlapalli, Krishnendu Ghosh and Arwinder Nagi)

AICHE 2012

(Paper ID number: 253849)

9/21/2012 1

Introduction to QbD- Approach and Advantages in Product and Process Development.

1. QbD approach provides Links b/w API, excipients and drug product
2. Design space, PAT , QRM and control strategy concepts
3. Regulatory Wins: Site and scale change flexibility
4. Faster approvals and minimal post approval issues
5. Visibility with global regulatory authorities

9/21/2012 2

Case Study-I

Problem Statement:

To explain the unexpected effect of a surfactant Poloxamer 188 on High drug Load, HSWG IR-tablet dissolution

Approach: Synergistic DoE effects, RSM , optimization and Multivariate data analysis

Objectives:

1. To explain tablet dissolution slowdown in HSWG IR-tablets with increase in poloxamer-188 in a combination DoE with the following design variables:
   a. Poloxamer amount
   b. Granulating water amount
   c. Binder addition rate

2. To elucidate the effects if any of the process variables on tablet dissolution
   Preliminary dissolution spec.: Q NLT 70% in 30 minutes in 0.1 N HCl

9/21/2012 3

What causes Poloxamer mediated dissolution slowdown?

- Poloxamer causes tablet DT slowdown mediated dissolution slowdown
- MOA of tablet DT slowdown:
  - Poloxamer mediated increased hydrophilicity of tablet matrix- exhausts early the disintegrant action of CPV
  - Poloxamer facilitates binder action of Povidone during HSWG by lowering granule IFT and ST of binder solution
    - » Evidence- Noisy /lumpy granulation (Run 2)
    - » High power consumption (Run 2)
  - Acts as a surrogate binder itself – Additional binding effect to Povidone.
  - Mechanism verified by no effect of poloxamer on CCS formulation

9/21/2012 4

Summary and Conclusions

- Poloxamer retards tablet dissolution while promoting granule dissolution by its effect on tablet DT
- MVDA complementary tool to DoE effects analyses explains above effect:
  - High poloxamer runs have coarser granulations,
  - Poloxamer does have a surrogate/inherent binding effect
- Synergistic use of DoE effects and MVDA provided essential clues into the MOA of poloxamer in tablets

9/21/ Publication: *"Quality by Design Case Study: Investigation of the role of poloxamer in immediate release tablets by experimental design and multivariate data analysis"*- AAPS Pharm. Sci. Tech. (2011), Vol 12, No.4, 1064-1076

Case Study II

Problem Statement:

HS Wet granulated product- Poor Blend flow ("Funnel flow and Rat-holing") observed at small scale process DoE batches. Blend flow was identified per QRA exercise and earlier experiments as being dependent on the following CPPs –

(i) Water Amount
(ii) Wet Massing time (WMT) and
(iii) Lubrication time (LT)

Objective/s:

1. Optimize blend flow using synergistic approach of DoE effects, RSM, optimization and MVDA
2. Develop a Control Strategy using the above 3 CPPs as design factors
3. Understand competing Blend flow and tablet dissolution

9/21/2012 6

Roadmap: Design to Knowledge Space by DOE, RSA and Optimization

DOE, and run experiments and obtain response variables → **DOE effect analysis to study significance of factors on responses** → **Response surface modeling and Optimization to establish design space** → **Run confirmatory batches within design space**

Run #	Lube Time (MIN)	WMT (MIN)	Water (G)	CTAB@ 30min	FFC
1a	3.0	1.0	268.7	88.6	7.4
2a	2.0	4.0	306.3	76.4	11.0
3a	3.0	5.0	268.7	83.6	7.6
4a	1.0	3.0	268.7	87.1	7.3
5a	5.0	3.0	268.7	85.0	6.0
6a	3.0	3.0	287.5	80.0	9.1
7a	4.0	4.0	306.3	83.7	4.8
8a	3.0	3.0	287.5	79.9	8.9
9a	3.0	3.0	287.5	81.5	8.2
10a	2.0	2.0	306.3	86.0	7.2
11a	4.0	2.0	306.3	86.9	7.2
12a	3.0	3.0	325.0	82.7	9.7
13a	3.0	3.0	250.0	82.7	10.5

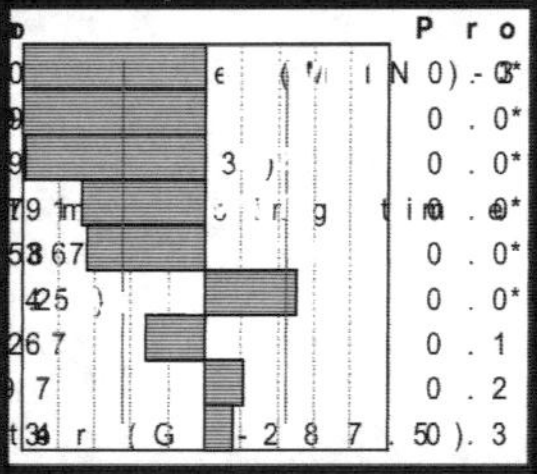

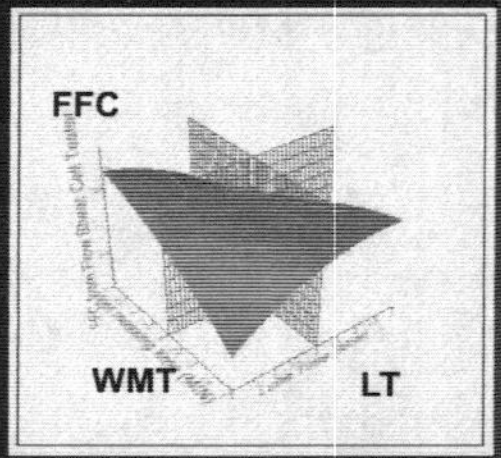

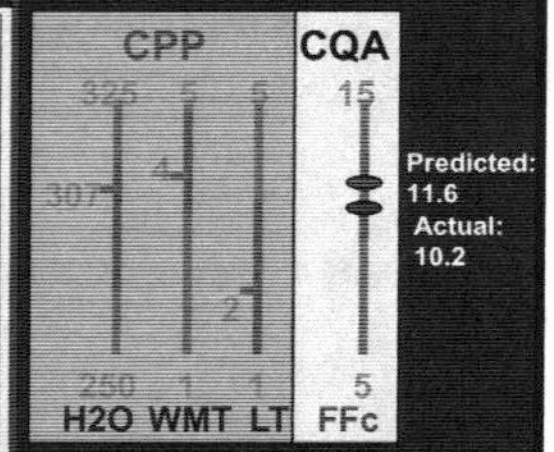

Knowledge space

DOE ranges of design variables were determined by trial runs (Sep 2008)

DOE Effect Analysis

Sort significance of modeling terms

Design space

An optimal region represented by a combination of variables that ensure desired CQAs

Confirmation

Run confirmatory batch using an optimal solution to verify design space (Dec 2008)

Summary and Conclusions

- Case studies exemplifies the application of QbD principles and tools to drug product and process development.
- Identified conditions/reasons for poor and good flow of tablet blend, optimized flow and established a design space
- Blend flow found to be improved in confirmatory runs by use of process parameters suggested by DoE effects analysis- High water and High WMT with low Lube times identified as being optimal for blend flow
- Control Strategy put in place for Blend flow, to ensure product always in DS region
- Dissolution data analysis- By MVDA, helped identify several variables responsible for dissolution differences and relationships b/w blend flow and dissolution. Compromise may have to be reached b/w the two.

Publication: *"Quality by Design Case Study: An Integrated Multivariate Approach to Drug Product and Process Development"*- Int. J. Pharm. (2009), doi:10.1016/j.ijpharm.2009.07.031

Acknowledgements

- L-Wyeth and Pfizer Drug product Co-dev team
- Chemometrician: Jun Huang
- Statisticians: Andrew McKeen and Brent Harrington
- Contract Manufacturing Facility.

(533g) Automation and Control of Drug-On-Demand Technology

Laura Hirshfield, Arun Giridhar, Gintaras V. Reklaitis, Michael T. Harris and Venkat Venkatasubramanian, Chemical Engineering, Purdue University, West Lafayette, IN

Stimulated by initiatives such as PAT and QbD, the pharmaceutical industry has been focusing on the development of innovative, efficient manufacturing methods. The concept of "mini-manufacturing" of drugs, as opposed to mass production of drugs, has shown promise in this area. We describe one such mini-manufacturing process, "drug-on-demand," which uses drop-on-demand printhead technology to deposit active pharmaceutical ingredients (API) onto edible substrates. We use a high precision positive displacement pump to deposit polymer-drug melts. Besides minimizing a need for mass production of all drugs, this method allows for the layering of different APIs and the creation of individual dosing forms, in which the amount of API can be varied depending on the patient.

In this work, we focus on the automation, control and Exceptional Events Management (EEM) strategy for the drug-on-demand pilot facility done as part of the Engineering Research Center for Structured Organic Particulate Systems. A proper control strategy is necessary so that the system can run at optimal setpoints and adjust in real-time, allowing for an efficient manufacturing process and a precise on-specification drug product. The control strategy for our "drug-on-demand" setup allows us to execute an automated, optimized, and controlled print cycle while closely monitoring drop size, drug morphology, and drop deposition pattern. The implementation of EEM allows for detection, diagnosis, and mitigation of abnormal events that occur outside of the control space to ensure a more efficient, productive process. In this session, we summarize our achievements and current research in this area.

QUANTITATIVE CHARACTERIZATION OF POWDER FLOW UNDER DYNAMIC CONDITIONS IN A ROTATING DRUM
[ID#275060]

John F Miller, Ph.D, GlaxoSmithKline, Research Triangle Park, NC

Introduction

Control of the flow properties of powders during one or more unit operations of a manufacturing process is an essential part of ensuring the process is well-controlled. Many assays exist that allow characterization of powder flow physical properties using modest amounts of material (10 to 100g) and in a non-production environment. However, most of these tests are conducted on powder samples that are static and/or constrained with respect to flow. Few assays exist that assess the powder under dynamic conditions. In addition, the equipment to perform such assays is often expensive and/or has limited flexibility due to design constraints imposed by the vendor. This work describes a rotary powder flow analysis instrument that addresses these issues and presents some practical examples of how it is a valuable complement to traditional methodology. This inexpensive method (less than $4,000) affords low measurement time (<10 minutes including data analysis), low sample requirement (3ml), and negligible material wastage. The use of inexpensive, disposable sample drums allows multiple samples to be prepared in parallel and kept for re-analysis if required. The examples are: comparison with ring shear cell testing for a range of common excipients used in the pharmaceutical industry, and comparison with granule size and dissolution kinetics as a function of milling parameters for a granulated blend of 80% API and 20% excipient. Related studies by other workers have been reported using analogous equipment but lacking the distinguishing characteristics of the instrument described herein [Refs. 1-6].

Instrument Design

The instrument's hardware configuration is similar to a discontinued commercial instrument *(Aero-Flow™, TSI Inc., Shoreview, MN)* and the currently available Revolution *(Mercury Scientific Inc., Newtown, CT)*. Each instrument employs a rotating drum containing the powder under test and uses optoelectronics to record changes of the location of the powder due to flow events such as avalanching (see Figure 1). Two light sources provide uniform illumination of the front of the rotating sample drum. The camera detects the motion of the drum and the sample within the drum. The output from the camera is transferred to a computer for subsequent analysis. The instrument is constructed from inexpensive, readily available consumer grade hardware and software, utilizes inexpensive disposable drums and requires typically less than 3ml of powder per measurement. The items are listed below. The total cost of ownership is approximately $3,700. However, most of the items were already available. The only additional items required were the illumination source, camera and development software thereby reducing the actual expenditure to less than $2,000.

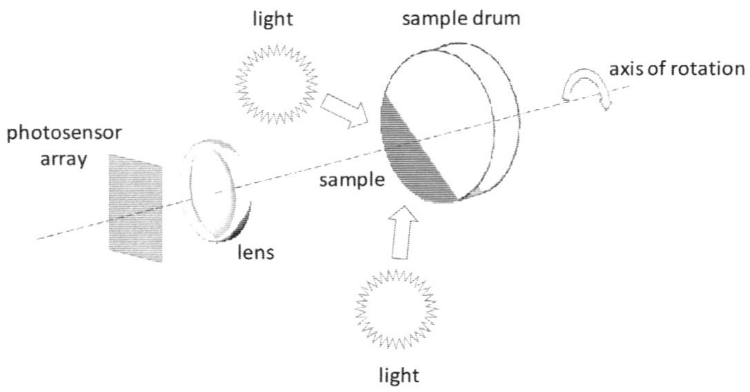

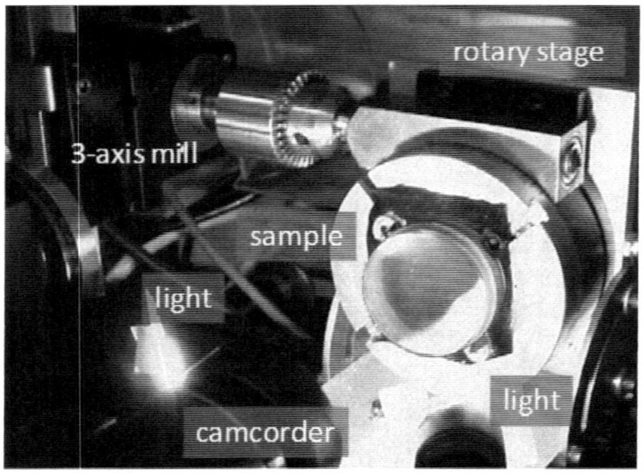

Figure 1. Schematic and photographic hardware configuration of rotary powder flow analyzer

The current instrument consists of the following components (including vendor and approximate 2012 price in US dollars):

- Sample drum – PS14 sample cup (45mm diameter x 14mm deep) ($1.25 each)
  (*Rotronic Instrument Corp.*, Hauppauge, NY)
- Sample drum holder – Sherline 3700 4" rotary table ($270)
  (*Sherline Products Inc.*, Vista, CA)
- Rotary table drive – Sherline 5000 3-axis mill ($650)
  (*Sherline Products Inc.*, Vista, CA)
- Illumination – SparkleLite Mini-Lite 350-2 Dual LED Light ($337)
  (*MK Digital Direct*, Chula Vista, CA)
- Camera – Sony HDR-HC9 camcorder ($1,099)
  (*Sony Corporation*, Tokyo, Japan)
- Computer – HP Compaq 8200 Elite personal computer with Microsoft Windows XP Professional XP operating system ($850)
  (*Hewlett-Packard Development Company*, Palo Alto, CA)
- Development software– Microsoft Visual Studio 2010 Professional ($500)
  (*Microsoft Corporation*, Richmond, WA)

Data Processing Algorithm

In addition to cost, the most important differences between the current instrument and the others are the data acquisition and processing algorithms employed. In particular, inter-frame analysis of the sequence of images received from the camera is used to provide valuable information regarding the movement of powder from one location within the drum to another. Calculation of traditional two-dimensional image moments [Ref. 7] allows the inter-frame changes to be represented with a few derived numerical responses. From these it is straightforward to calculate estimates of such properties as flow event time, mass, trajectory and so on. In addition to the novel processing algorithms, the instrument allows for a persistent record of the video information and subsequent re-analysis with alternative algorithms.

The specific algorithm used for the studies described herein is illustrated in Figure 2. Two adjacent images 1 and 2 from the camera are acquired by the computer. These unprocessed video frames are converted to grayscale images and reduced in resolution to 320 x 240 pixels. Higher resolutions can be used but this significantly increases the amount of data processing required. The images are subsequently converted to a binary format – i.e., the images contain only black or white pixels. A circular vignette is also applied at this stage to keep only the sample drum region of the image. The processed version of frame 1 is defined as the "full" image. A new image is created using white pixels that only occur in frame 1. This represents the powder that is "lost" from frame 1. A similar image is created for powder that is "gained" by frame 2. These are "delta" images.

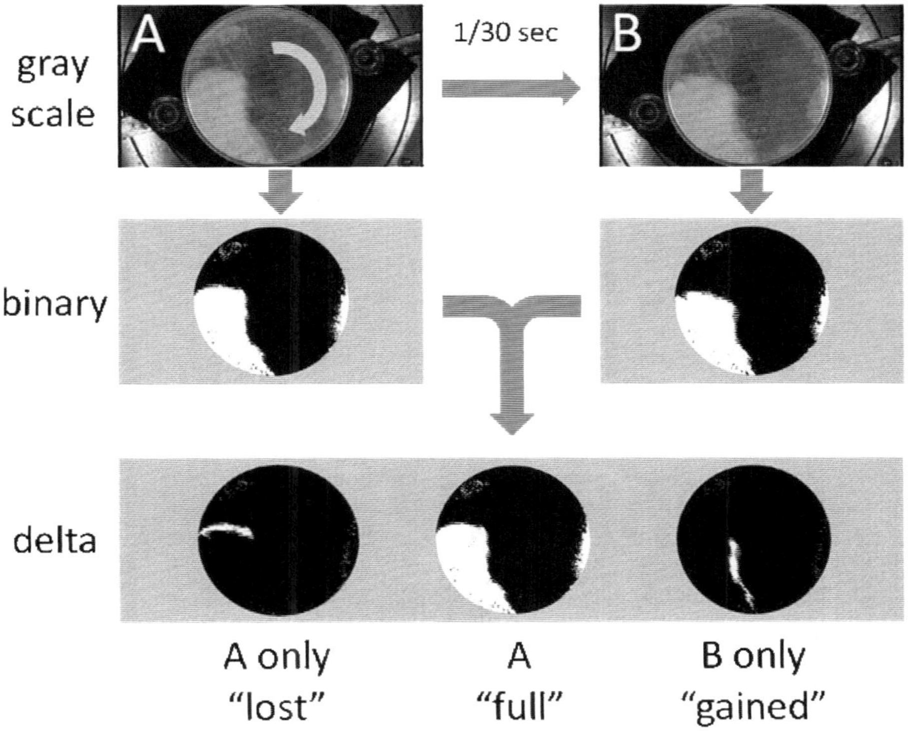

Figure 2. Illustrative representation of data processing algorithm

Hence, for each pair of adjacent images, there are three derived images each containing white two-dimensional objects set on a black background. Well-established simple image processing algorithms [Ref. 7] allow a variety of descriptors to be calculated for each image. The simpler of these include the total area of the object (the cross-sectional area of powder) and the center of mass (the location relative to the center of the drum). In other words, it is possible to estimate how much powder moves and where from with time. Generally, an experiment analyzes 3ml of powder in the drum that is rotating at a fixed rate between 0.5 and 15rpm for five minutes. This corresponds to approximately 9,000 images. The image processing and descriptor calculations are performed using bespoke software and the results exported to data processing software. For an experimental duration of 5 minutes, the image processing and calculations require approximately 30 seconds to complete. This permits analysis to be performed as images are transferred from the camera (i.e., in so-called real-time). However, it is preferable to perform the analysis after all the images have been collected since this allows pre-processing parameters to be optimized on an experiment-by-experiment basis. Further data processing may be performed using a variety of commercial software applications. Typically, the means and standard deviations for each descriptor are calculated.

From experience, two specific image descriptors are advantageous: the height of the center of mass of the object in the "full" image (referred to as *height (full)*) and the area of the object in the "lost" image (referred to as *area (lost)*). Physically, these provide a measure of the chaotic behavior of the movement of powder. Figure 3 shows an example of these two descriptors.

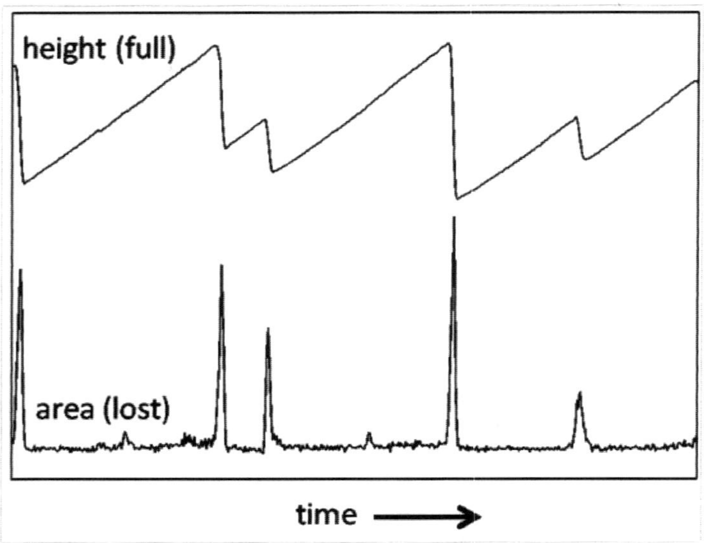

Figure 3. Example of two image descriptors *height (full)* and *area (lost)* as a function of time

Excipient Study

Twelve common pharmaceutical excipients were studied using the rotary powder flow analyzer and the results compared to those obtained using different techniques. The excipients are listed below:

1. Pearlitol 100SD	**2.** Avicel PH200	**3.** Pearlitol 200SD	**4.** Pearlitol 160C
5. Avicel PH101	**6.** Avicel PH102	**7.** SuperTab 11SD	**8.** Pharmatose 200M
9. Avicel PH105	**10.**SuperTab 21AN	**11.**Avicel PH102	**12.**Starch 1500

In each case, 3ml of powder was analyzed at a rotational speed of 0.5rpm for 5 minutes. Figure 4 shows a montage of single images obtained during the measurement for each excipient. The numbers correspond to the list of excipients above. Differences in appearance are evident particularly for both 8 and 9 when compared to the others.

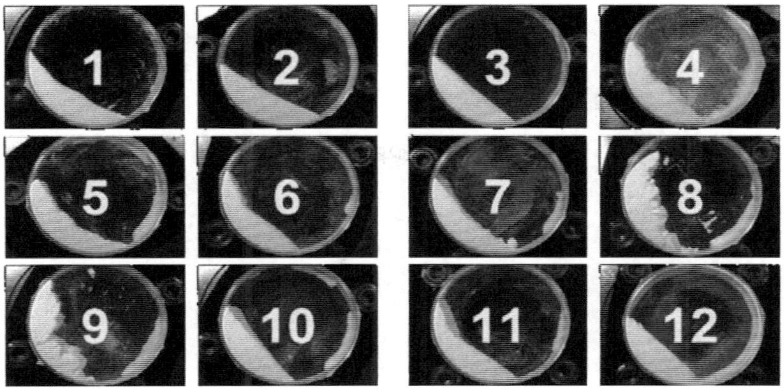

Figure 4. Single images obtained during the measurement for each excipient (refer to main text for additional information)

Comparison with ring shear cell test

The results of the analysis for each excipient show very good correlation with those for ring shear cell testing (*Model PFT, Brookfield Engineering Laboratories, Inc. Middleboro, MA*). In particular, the standard deviation of the *area (lost)* descriptor correlates well with a number of responses from the shear cell tests including the shear stress to normal stress ratio, internal friction angle, and flow function (expressed as a reciprocal value). The results of the comparisons are shown below. Similar, slightly weaker correlations are found with the *height (full)* descriptor (not shown). The data markers are coded according to the corresponding excipient family (e.g., microcrystalline cellulose, mannitol etc.). The dashed lines represent the 95% confidence curves of the linear model of the experimental data.

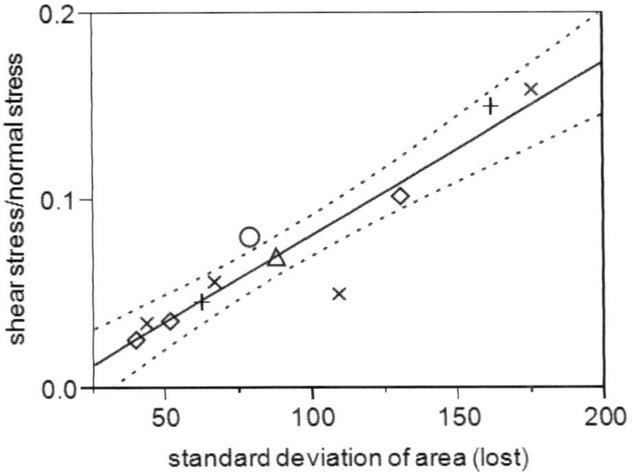

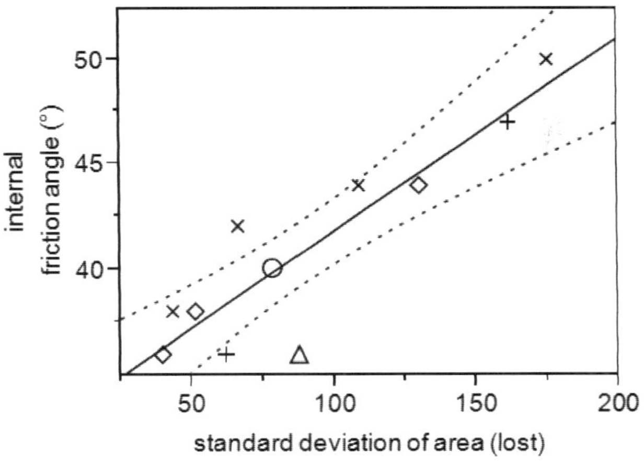

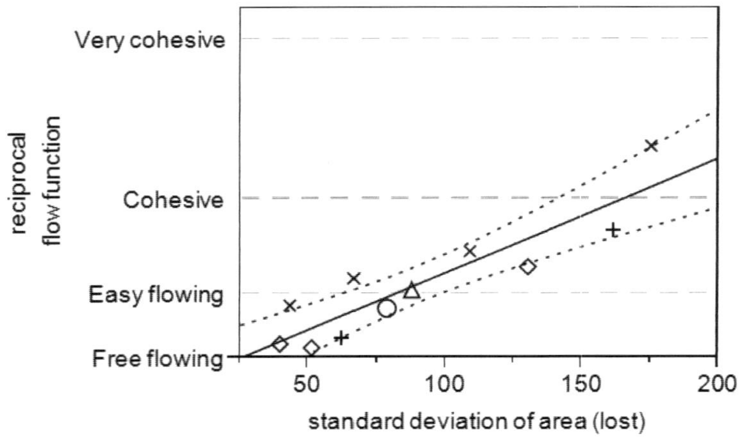

Figure 4. Correlation between rotary powder flow measurements (standard deviation of *area (lost)* descriptor (arbitrary units)) and various ring shear cell responses (*top left* – shear stress/normal stress, *top right* – internal friction angle, *bottom* – flow function (expressed as reciprocal value))

Comparison with particle size via laser diffraction

Good correlation is observed with the particle size fraction less than 50 micron as determined by laser diffraction *(Mastersizer 2000, Malvern Instruments Ltd, Malvern, UK*) (Figure 5).

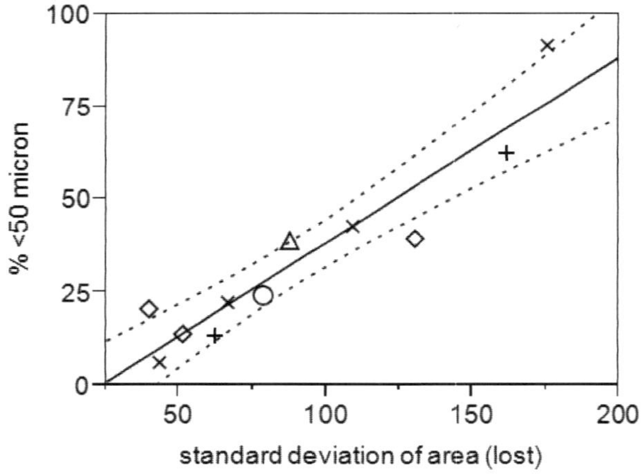

Figure 5. Correlation between particle size and rotary powder flow measurements (see text for explanation)

Wet Granulation Study

An experimental study was conducted to look at the effect of milling speed and screen size on the final granule size determined via vibratory sieve analysis *(Model AS 200, Retsch Technology GmbH, Haan, Germany)* and dissolution kinetics (*Zymark Multidose Workstation, SOTAX AG, Allschwil, Switzerland*) for a blend of an API and an excipient (80:20 %w/w, respectively). Samples were tested using the rotary powder flow analyzer (3ml sample, 15rpm rotational speed, 1 minute duration) and the results compared to the particle size and dissolution data. Two milling speeds (1800rpm and 3000rpm) and three screen sizes (0.025, 0.032 and 0.045 inches) were studied (*CONSIGMA™ Continuous Granulation and Drying System, GEA Process Engineering nv, Wommelgem, Belgium*). For each of the three tests, the following responses were found to correlate well with each other: standard deviation of *area (lost)*, dissolved amount after 45 minutes and the fraction less than 125 micron. The results are shown in Figure 6 for milling speed 1800rpm (left) and 3000rpm (right). To aid comparison, the results from each technique are normalized such that the minimum-maximum range is scaled from zero to one and superimposed on the same graph. The marker-line style coding is: open diamonds-dashdotted = *area (lost)*, pluses-dotted = fraction less than 125 micron, and open circles-dashed = dissolved amount after 45 minutes. The lines are provided as a visual guide only.

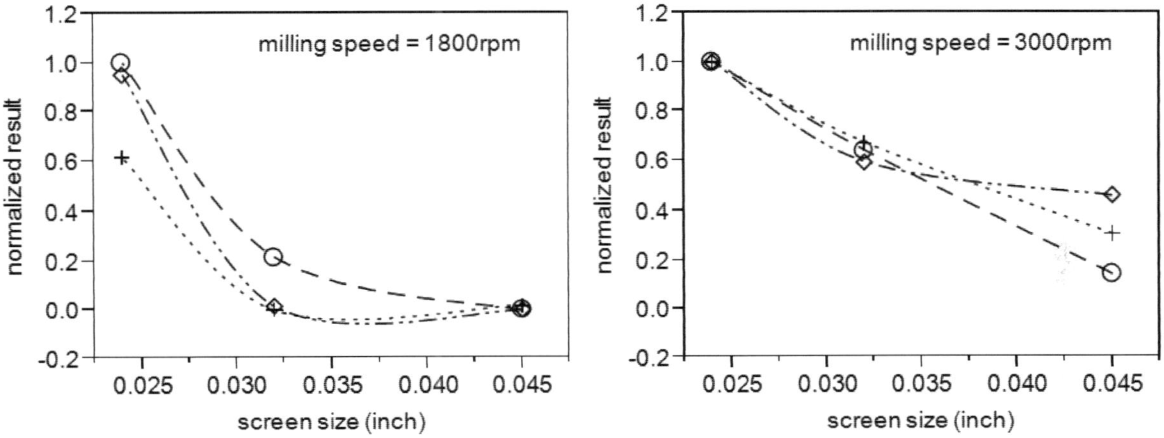

Figure 6. Normalized responses as a function of screen size at two different milling speeds as determined using three different measurement techniques (see text for details)

The rotary powder flow analysis results follow the same general trends as for the dissolution and particle size data.

Discussion

Previous studies of the flow of pharmaceutical powders within a rotating drum have been reported particularly using commercial implementations of the technique [Refs. 2, 4, 5, 6]. The implementation described herein is an extension of these instruments and confers a number of important advantages as well as providing a complementary method for powder flow characterization using methods such as shear cell testing. The benefits lend themselves to application in early development phases of powder-based formulations.

The most significant unique feature of the current instrument is the inter-frame image processing as described earlier. Such processing, together with the equivalent for intra-frame processing, permits many different descriptors to be calculated with minimal computational effort. Indeed, as has been explained, two simple descriptors can adequately describe the dynamic changes occurring with the sample as it rotates. Furthermore, these descriptors can be explained readily in terms of basic physical phenomena. The first descriptor (*height (full)*) simply indicates how high the sample can rotate before powder flow occurs and the second (*area (lost)*) describes how much powder moves as a result. The standard deviations of these values indicate the variability of the flow events. Powders that flow easily will tend to exhibit a "rolling" flow regime [Ref. 5] whereby the sample height remains constant and low, and powder flow is steady. Hence, the standard deviations for both descriptors will be small. For cohesive powders, flow will tend to exhibit "cataracting" behavior that will be increasingly more erratic with increasing cohesivity. Such powders can rotate further up the side of the drum before gravitational forces overcome internal forces and a significant movement of powder occurs. These will be more so for less dense powders, too. Hence, the standard deviations for both descriptors will be higher. This physical description is borne out by the ring shear test results for the excipient study. Specifically, it is shown in Figure 5 that the shear stress/normal stress ratio and, separately, the internal friction angle exhibit a strong, linear correlation with the standard deviation of *area (lost)*. i.e., the more chaotic the powder flow in the rotating drum, the higher the internal strength of the powder. In the case of the wet granulation study, the same physical explanation helps describe the observed behavior. Specifically, as the particle size decreases, the powder flow in the drum becomes more erratic, as would be expected.

Previous published work reports the powder flow behavior in a graphical manner in the form of a "strange attractor" plot. In essence, such a plot indicates the variability of the time interval between significant powder movement events in the rotating drum. The disadvantages to this approach are that, generally, the data are only presented pictorially and most of the raw data are rejected. In contrast, the new method uses all the images received from the camera. Additionally, the same data can be processed in a similar manner to the previous implementations if necessary but, based on experience, this offers no advantage. Indeed, the significantly reduced number of data points used to calculate the final output from the instrument decreases the signal-to-noise ratio.

The calculation of the image descriptors is computationally trivial. For each image pair that is analyzed, the results are written to a simple text format file (typically a comma-separated value file). Once the entire set of images from an experiment has been analyzed, the file is imported to a data processing software application such as *Microsoft Office Excel* and the standard deviations of the descriptors thereby calculated, together with any other numerical processing as appropriate. The total time required from receipt of a bulk powder sample to generation of the final numerical result is less than ~10 minutes.

The selection of the sample drum for the instrument confers another advantage compared to the previous versions of the technique. Those both require dedicated, proprietary drums. Hence, only one sample can be prepared and tested at a time. The drum needs to be cleaned between samples and a given sample cannot be tested again. For both instruments, typically 20 – 100ml sample are required. For use in early pharmaceutical development, this can be a prohibitively large amount. The new instrument uses inexpensive, low internal volume, disposable drums. All of the measurements presented in this article used 3ml of powder. Multiple samples can be prepared ahead of measurement. For example, during a milling campaign, 3ml samples could be removed and put directly into separate sample drums. Following analysis, the samples and drums can be stored and re-tested if necessary.

Conclusions

An inexpensive instrument has been developed that can provide quantitative information regarding the flow of powders rotating unconstrained in a drum. Correlation has been demonstrated between this rotary powder flow analyzer and a number of other physical property measurement techniques, namely powder ring shear testing, particle sizing via laser diffraction and sieve analysis, and dissolution testing. The method is inexpensive to implement (<$4,000), affords low measurement time (<10 minutes including data analysis), low sample requirement (3ml), and negligible material wastage. The use of inexpensive, disposable sample drums allows multiple samples to be prepared in parallel at locations remote from the instrument and kept for re-analysis if required.

Acknowledgments

The author thanks the following people within GlaxoSmithKline for the contribution of experimental data for the ring shear cell, sieve and dissolution analyses: James M Kennady, Stacey T Long and Chris C Cook. William L Roberts is thanked for his expert advice regarding the incorporation of the milling machine and rotary table into the overall hardware configuration.

References

1. Brian H. Kaye, Julie Gratton-Liimatainen and Neil Faddis (1995), "Studying the Avalanching Behaviour of a Powder in a Rotating Disc", *Particle & Particle Systems Characterization,* **12**(5), pp. 216-266.

2. Michael K. Taylor, Jeri Ginsburg, Anthony J. Hickey and Ferdous Gheyas, "Composite Method to Quantify Powder Flow as a Screening Method in Early Tablet or Capsule Formulation Development" (2000), *AAPS PharmSciTech,* **1**(3) article 18.

3. Riccardo Bonfichi, Giulia Cloralio and Andrea Rainoldi, "Dynamic avalanching accurately assesses flowability and quality" (2009), *Pharmaceutical Technology*, July.

4. Venkateshwar Rao Nalluri and Martin Kuentz, "Flowability characterisation of drug-excipient blends using a novel powder avalanching method" (2010), *European Journal of Pharmaceutics and Biopharmaceutics,* **74**, pp. 388-396.

5. Bruno C. Hancock, Kim E. Vukovinsky, Barry Brolley, Ian Grimsey, David Hedden, Angela Olsofsky and Rebecca A. Doherty, "Development of a robust procedure for assessing powder flow using a commercial avalanche testing instrument" (2004), *Journal of Pharmaceutical and Biomedical Analysis,* **35**, pp. 979-990.

6. Francois Lavoie, Louis Cartilier and Roch Thibert, "New Methods Characterizing Avalanche Behavior to Determine Powder Flow" (2002), *"Pharmaceutical Research"*, **19**(6), pp. 887-893

7. Ming-Kuei Hu, "Visual pattern recognition by moment invariants" (1962), *"Information Theory, IRE Transactions on"*, **8**(2), pp. 179-187

(645b) Impact of Formulation Properties On Lubrication of Common Tableting Excipients

Benjamin J. Baranowski and **Joseph Kushner IV**, Drug Product Design, Pfizer Worldwide Research and Development, Groton, CT

A recently-developed lubrication blending model identified formulation-dependent parameters for lubrication sensitivity, β, and lubrication rate constant, γ. The original study examined binary mixtures of common tableting excipients. Therefore, there is a need to determine if the lubrication model is applicable for a wider range of solid oral pharmaceutical tablet excipients by observing the impact of formulation properties (i.e. lubricant level, particle size and morphology) on the bulk specific volume and tablet tensile strength of lubricated excipients. 18 excipients were examined this study. All 18 excipients were evaluated for lubrication sensitivity analysis at a particle sieve cut range of 90-125 microns and 1 wt-% magnesium stearate. 6 excipients were selected for particle size and level of lubrication study and examined using a 2^2 experimental design with center point. The previously-developed lubrication model was found to adequately describe the impact of lubrication on tensile strength and bulk specific volume of a wide range of excipients. γ was similar for bulk specific volume and tablet tensile strength changes for a wide range of excipients, while b exhibited less correlation. Excipient particles with round shapes and smooth surfaces appear to have higher γ values (e.g. spray-dried lactose, $\gamma \sim 10^{-2}$ decimeters^{-1}), and, therefore, are more quickly lubricated by magnesium stearate, than rough, irregularly shaped particles (e.g. crospovidone, $\gamma \sim 10^{-5}$ decimeters^{-1}). γ may also depend on the cross-sectional area of the formulation particles and on the level of lubricant in the formulation. The results suggest that material properties of the excipients and the lubrication level in the formulation are a significant contributor to the sensitivity of a formulation to lubrication effects. Further, the results suggest that formulations can be designed to provide reduced sensitivity to adverse effects associated with lubrication, thereby improving the overall robustness and quality of the drug product.

(645c) Formulation Development of Hot Melt Extrudate to Stabilize Amorphous API

Sandeep B. Kedia, Chemical Development, GlaxoSmithKline Inc, Research Triangle Park, NC

Hot Melt Extrusion: Identification and optimization of key excipients utilizing high-throughput screening methodology.

By utilizing various tool kits available in the chemical engineering toolbox, a case study would be presented on formulation development in an accelerate timeframe for a marketed drug product. A new high throughput screening methodology will be presented, to screen various excipients, in order to quickly identify the optimum extrusion composition, using a small scale (1-mL) vials, based on blend uniformity, Tg and heat capacity for the various mixture of interest. To optimize the throughput requirements for manufacturing scale, an ordinary differential equation would be used to solve the required temperature profile of the extrudate in order to achieve desired properties, such as, swelling of the strand, tackiness to cut, elastic and brittle fracture of the strand. Also, work on new hot face pelletization technique would be presented for containment issues for a highly potent (OHC-5 category) API, to achieve the desired pellet distribution in order to prepare a uniform blend for tablet development.

(645d) Agglomerate Characterization in Pharmaceutical Systems

Dushyant Shekhawat, Analytical Sciences Research & Development, Eli Lilly & Company, Indianapolis, IN

Particle agglomerates (large particles comprised of small primary particles) can be formed in various unit operations in the pharmaceutical industry. Physical properties (i.e., size, shape, and strength) and amount of these agglomerates can affect the critical quality attributes of drug substance and drug product. This presentation will focus on various approaches that can be used to detect and characterize agglomerates when present in a bulk power, including the benefits and limitations of each analytical technique. Additionally, information will be presented on small scale studies that were designed to provide a thermodynamic understanding of agglomerate formation. Results from these small scale studies were used to design an optimized packing strategy for the powder to avoid the formation of detrimental agglomerates.

(645e) Development of API Target Attribute Profile

Weili Yu, Drug Product Design, Pfizer Inc., Groton, CT and Subodh Deshmukh, Drug Product Design, Pfizer Inc, Groton, CT

API physical attributes play an important role in drug product design and development. These attributes often have a direct impact on the performance, stability and manufacturability of drug products of various dosage forms. Identification and characterization of these critical attributes is important for successful design and manufacture of API optimized for the targeted dosage form. While the importance of API properties is commonly recognized in the industry, the integration of API and drug product design during the development process is not nearly as common. The API Target Attribute Profile (TAP) is a venue that facilitates discussions between the API designers (chemical development groups) and the drug product designers (formulation groups) and ensures that the key API physical properties and acceptable ranges are agreed upon and optimized for the dosage form. Case studies will be presented to demonstrate the process for establishing the TAP along with advanced material characterization techniques used to support the process.

(645f) Application of Materials Science Approaches to Facilitate Drug Substance and Product Processing

Thomas Borchardt, Global Pharmaceutical Research & Development, Abbott Laboratories, North Chicago, IL and Raimundo Ho, Process Research & Development, Abbott Laboratories, North Chicago, IL

Application of Materials Science Approaches to Facilitate Drug Substance and Product Processing

Raimundo Ho and Thomas B Borchardt

Abbott Laboratories, Global Pharmaceutical R&D

An important aspect of the development of drug substance manufacturing processes is the identification of the properties critical to formulation processing. Historically, the emphasis has been on particle size with less attention to other physical properties such as particle shape and surface properties. The emergence of new characterization technologies in the pharmaceutical industry has broadened the scope of properties that can be evaluated, but simultaneously expands the amount of data available for interpretation. From a practical point of view, modeling approaches that facilitate a quick determination of the key attributes that impact the processing behavior of the drug substance can drive efficiency in property characterization. A chemometric approach has been developed to identify critical physical property parameters impacting flow function coefficients (FFCs) of bulk powders. A partial least squares (PLS) model was established to correlate metrices of particle physical property characteristics with powder FFCs, with the aim of maximizing model predictability (Q^2) whilst minimizing the number of latent factors and root mean square error of the response. The model indicates that bulk powder flow is not only affected by particle size distribution, but may also be significantly affected by the shape distribution, span of the distribution, and surface energetics. An understanding of the fundamental physical property characteristics will drive efficiency in characterization and aid the design of isolation processes to target particle attributes for downstream formulation and processing needs.

On The Modeling of Oral Drug Delivery

Naresh Pavurala, Luke E.K. Achenie, Virginia Polytechnic Institute and State University, Blacksburg, Virginia, USA

Abstract

A drug release mechanism is combined with a Compartmental Absorption and Transit (CAT) model within a pharmacokinetic framework to predict orally administered drug release profile and pharmacokinetic parameters such as peak plasma concentration (C_{max}), area under the curve (AUC) and bioavailability. The pharmacokinetic framework is coupled with an optimization model to evaluate optimal values of key parameters such as tablet size for a target plasma concentration profile. The model plasma profile showed good agreement with the clinical profile obtained from literature. The pharmacokinetic parameter values evaluated were within 10% of deviation from the clinical values.

Keywords: oral drug delivery, ACAT model, dissolution model, pharmaco-kinetic simulation, optimization

Introduction

Drug delivery can be described as application of chemical and biological principles to control the in vivo temporal and spatial location of drug molecules for clinical benefit. When a drug is administered, only a small fraction of the dose actually reaches the relevant receptors or sites of action, and most of the dose is wasted either by being taken up into the wrong tissue, removed from right tissue too quickly, or destroyed en route before arrival. The research in drug delivery is to tackle these issues in order to maximize drug activity and minimize side effects.

Oral administration presents a series of attractive advantages towards other drug delivery. These advantages are particularly relevant for the treatment of pediatric patients and include the avoidance of pain and discomfort associated with injections and the elimination of possible infections caused by inappropriate use or reuse of needles. Moreover, oral formulations are less expensive to produce, because they do not need to be manufactured under sterile conditions (Salama, Eddington et al. 2006). In addition, a growing body of data suggests that for certain polypeptides such as insulin; the oral delivery route is more physiological (Hoffman A 1997). It is estimated that 90% of all medicines are oral formulations and oral drug delivery systems comprise more than half the drug delivery market. However, biopharmaceutical issues such as physicochemical requirements of the drug and physiological conditions make oral delivery one of the most challenging routes.

Dissolution and absorption of drugs from the gastrointestinal (GI) tract is a very complex process involving multiple steps including drug disintegration and dissolution, degradation, gastric emptying intestinal transit, intestinal permeation and transport, intestinal and hepatic metabolism (Yu, Lipka et al. 1996; Martinez and Rilyn 2002; Huang, Lee et al. 2009). It is influenced by many factors that not only vary among various compounds and formulations but also between different regions in the

GI tract with respect to rate and extent of absorption. Absorption properties also vary from subject to subject and from time to time. The GI tract offers a large absorption surface and nearly one third of the blood from cardiac output flows through the gastrointestinal organs making it a very good site for drug absorption (Gabor, Fillafer et al. 2010).

The drug discovery process involves the use of considerable resources to analyze the pharmacokinetic properties of the potential drug candidates. Pharmacokinetic modeling aids in optimizing the use of such resources in the *in vivo* testing by predicting the pharmacokinetic behavior of the drug candidates. These models also have the potential to filter out the most promising drug candidates for development and separate them from the ones with low probability of success (van de Waterbeemd and Gifford 2003). The best way to develop a pharmacokinetic model is by incorporating the processes of absorption, distribution, metabolism and excretion (ADME) into one global model. Mechanistic models strive to represent the physiological and pharmacological processes as accurately as possible using first principles approach. They incorporate our current understanding of the structure of the physiological system and the biological processes involved such as drug dissolution, degradation, gastric emptying, intestinal transit, intestinal permeation and transport, intestinal metabolism and hepatic metabolism (Marshall, Macintyre et al. 2006).

There are different kinds of mechanistic models such as dispersion models and compartmental models. There are dispersion models that define the GI tract as a single tube with spatially varying properties (pH, surface area, etc.) along the tube (Ni, Ho et al. 1980) in contrast to compartmental models which represent the small intestine as a series of compartments with linear transfer kinetics, with uniform concentration in each compartment. The movement of drugs through the intestine was modeled by Willmann and co-workers (Willmann, Edginton et al. 2007) through the introduction of an intestinal transit function (defined as the fractional drug dose in a specific GI segment at a given time) into the dispersion model. The model accounted for effects of solubility and permeability on the drug absorption and successfully simulated the fraction of dose absorbed for a wide variety of drugs in rats (Willmann, Schmitt et al. 2003), humans (Willmann, Schmitt et al. 2004) and monkeys (Willmann, Edginton et al. 2007). It was observed that the dispersion model predicted well for passively absorbed drugs. However, assumptions such as negligible first-pass effect and passive intestinal permeability do not appear justified. A typical CAT model (Yu, Lipka et al. 1996; Yu and Amidon 1999) divides the GI tract into three segments namely, stomach, small intestine and colon, while the transit flow in the small intestine is separated into seven compartments. The model considers simultaneous small intestinal transit flow and drug absorption. In addition the model assumes instantaneous dissolution, linear transfer kinetics and equal transit times in all the compartments, very low absorption in stomach and colon. The dependence of the fraction of dose absorbed on the effective permeability of various drugs with different absorption characteristics was well captured by the model (Yu and Amidon 1999). This model serves as a basis for most of the compartment models being used in pharmacokinetics.

Dissolution Model

The Dissolution Model (Narasimhan and Peppas 1997) explains the release of a water soluble, crystalline drug enclosed within a swellable, hydrophilic glassy polymer, placed in contact with water or a biological fluid. It is a moving boundary problem formulated using a system of partial differential equations (PDE's) and ordinary differential equations (ODE's) which describe the diffusion of (i) water into the system; (ii) drug out of the system; (iii) polymer chain disentanglement through the boundary layer; (iv) the left and right moving boundaries.

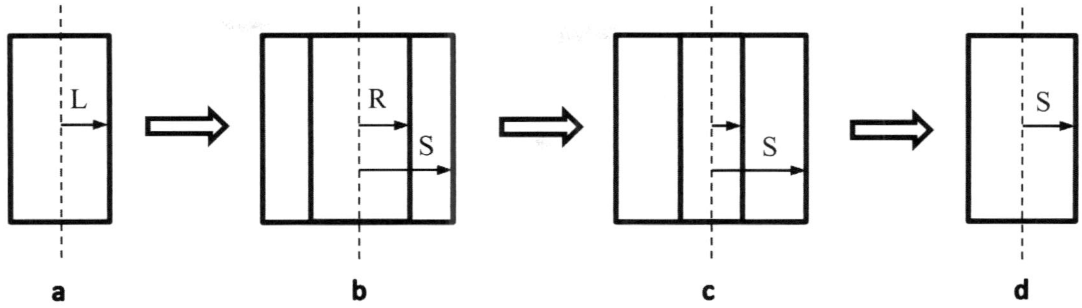

Fig. 1: One-dimensional solvent diffusion and polymer dissolution process(Narasimhan and Peppas 1997)

ACAT Model

In this work, an Advanced Compartmental and Transit (ACAT) model (Agoram, Woltosz et al. 2001) was used as a basis for developing the pharmacokinetic model. It includes linear transfer kinetics, six plausible states of the drug component (unreleased, undissolved, dissolved, degraded, metabolized and absorbed), nine compartments (stomach, seven segments of the small intestine, and colon), and three states of the excreted material (unreleased, undissolved and dissolved). Physiological factors such as gastric emptying, intestinal transit rate, first-pass metabolism, luminal transport and physicochemical factors (e.g. pKa, solubility, particle size, particle density, permeability) and dosage factors including dosage form and dose are taken into account in predicting the oral drug absorption (Huang, Lee et al. 2009). The ACAT model does not account for the process of dissolution as in (Narasimhan and Peppas 1997) where the disentanglement of the polymer is also taken into account along with the drug dissolution, clearly explaining the dissolution process.

Modeling Strategy

The pharmacokinetic model was developed by combining the dissolution model with the ACAT model. A schematic of the developed model is given in Fig. 2. The model employs linear transfer kinetics with two plausible states of the drug component (dissolved and absorbed) as shown in Fig. 2. The first two layers in the ACAT model representing the unreleased and released states of the drug component are replaced by input from the dissolution model as shown in Fig. 2. The basic approach is that the amount of dissolved drug, which enters each of the compartments in the ACAT model, is evaluated using the dissolution model. The latter assumes that the drug tablet is cylindrical and that dissolution occurs in the radial direction. Hence, the drug feed into each of the compartments is evaluated using the density of the drug, thickness of the tablet, fraction of drug in the tablet and the compartment volume obtained from Bolger et al. (Bolger 2009). The drug feed determines the plasma concentration profile of the drug. Note that the drug dissolves gradually as it travels down the GI tract. The pharmacokinetic model is embedded in an optimization framework, in order to evaluate key system parameters such as the optimal thickness of the tablet such that the predicted plasma concentration profile matches the clinical plasma concentration profile as closely as possible.

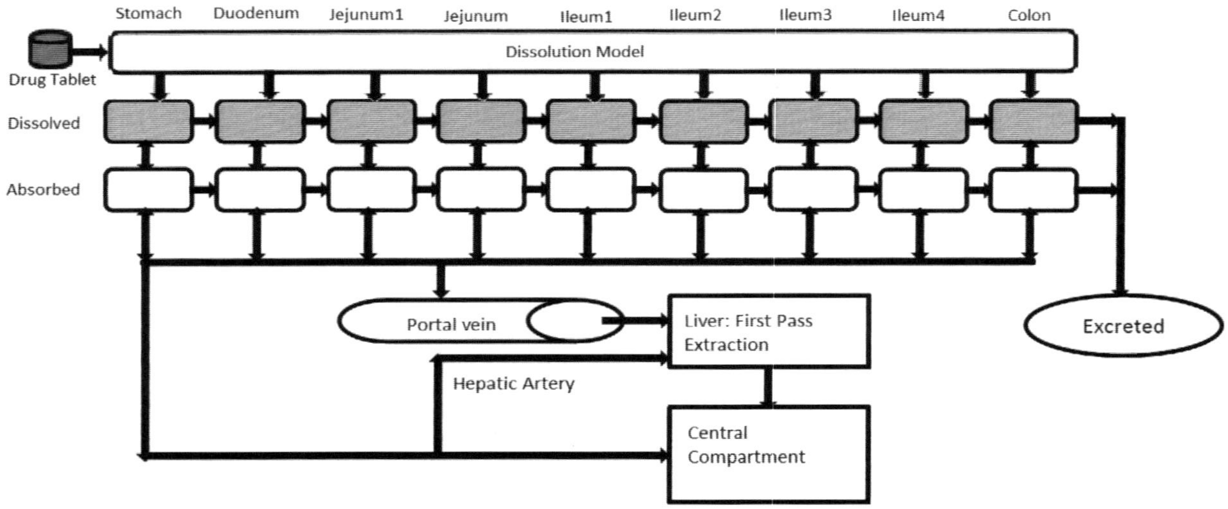

Fig 2: Schematic of the developed pharmacokinetic model

Results and Discussion

In this paper, Cimetidine is used as a model drug to test the pharmaco-kinetic model. Cimetidine is an antagonist of the Histamine H_2-receptor that inhibits the secretion of gastric glands. The drug is available in a variety of oral dosage forms for the treatment of duodenal and gastric ulcers (Somogyi and Gugler 1983). Plasma concentration-time profiles of Cimetidine in 12 healthy patients following oral administration of a commercially available immediate release formulation (Tagamet) as well as three extended release formulations was presented in Jantratid *et al*(Jantratid, Prakongpan et al. 2006). Bioavailability and pharmacokinetics of cimetidine in 10 healthy volunteers after administration of oral doses (100,400, and 800 mg) is studied in Grahnen *et al* (Grahnén, Bahr et al. 1979).

A clinical plasma concentration profile for cimetidine was obtained from Jantratid et al (Jantratid, Prakongpan et al. 2006) in which, a 400 mg cimetidine tablet was used as the initial dosage in the experiment which contained 40% of drug and 7.5% of polymer. A plasma profile with peak plasma concentration (C_{max}) of 2100 ng/ml and a bioavailability of 64% (well within the range of clinical results) was considered for the model. The optimization was started with an initial dosage of 190 mg (considering just the mass of drug and polymer). The density of cimetidine is taken as 1.4 g/cc and the effective human permeability as 0.287×10^{-4} cm/s. The total time of simulation is taken as 12 hours. The transit times, volumes and radii of each compartment were taken from (Table I) in (Bolger 2009). The thickness of the tablet was varied during the optimization.

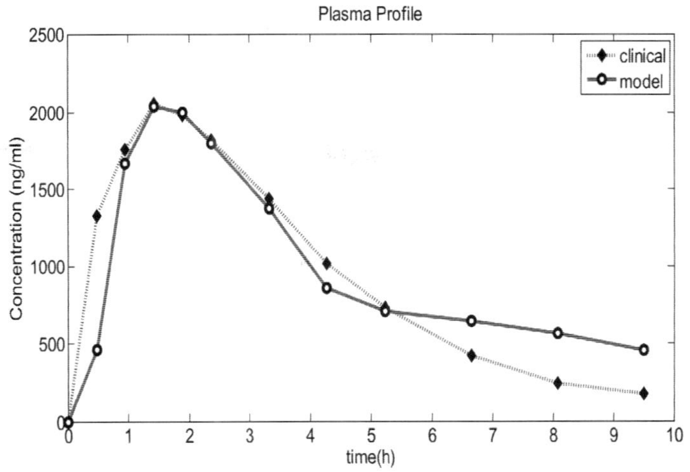

Fig. 3: Predicted and clinical plasma concentration profiles of cimetidine oral dose. The dotted line represents the clinical plasma concentration profile obtained from literature and the solid line represents the predicted plasma concentration profile using the pharmacokinetic model.

The optimal thickness of the tablet was obtained as 1.034 mm. The value of Cmax was obtained as 2041.2 ng/ml, which is comparable to the clinical value of 2055 ng/ml. The area under the curve (AUC) values obtained for the clinical and predicted profiles in Fig. 3 are 8954.6 ng.hr/ml and 9212 ng.hr/ml respectively. In both profiles, the peak plasma concentration (C_{max}) was achieved within 1.5 – 2 hours. The bioavailability was evaluated as 65.23% which is very close to the bioavailability of 64% obtained from literature..

Conclusions

Coupling ideas from the ACAT model and the dissolution model, a complete pharmacokinetic model is developed. The predicted plasma concentration profile matches well with the clinical profile, except that after 5 hours, the curves diverge a little. We see a gradual decrease in concentration of the drug in the predicted curve but more steep decrease in the clinical curve. This may be due to a higher rate of first-pass elimination than that accounted in the model.The values of the pharmacokinetic parameters such as C_{max}, AUC calculated from the model are comparable to clinical results. The advantage of the suggested approach is that it can be used to test hypothesis about the mechanism involved in drug delivery into the blood circulatory system. The model also evaluates the optimal thickness for the drug tablet in order to obtain the required plasma concentration profile. We are currently working on a triphasic release profile comprising of an initial burst, lag phase and final release of the drug. This study aids in controlling the different phases of drug release through optimization. The advantage of the suggested framework is that it can be used to test hypothesis about the mechanism involved in drug delivery into the blood circulatory system. We expect to employ the composite model to hypothesize the optimal diffusion coefficient (and other physico-chemical properties) that a new drug needs in order to obtain a target drug release profile. This is expected to be useful for drug design.

References

1. Agoram, B., W. S. Woltosz, et al. (2001). "Predicting the impact of physiological and biochemical processes on oral drug bioavailability." Advanced Drug Delivery Reviews **50, Supplement 1**(0): S41-S67.

2. Bolger, M. B. (2009). "Simulations of the Nonlinear Dose Dependence for Substrates of Influx and Efflux Transporters in the Human Intestine." The AAPS Journal **11**(2): 353-363.

3. Gabor, F., C. Fillafer, et al. (2010). Improving Oral Delivery

4. Drug Delivery. M. Schäfer-Korting, Springer Berlin Heidelberg. **197:** 345-398.

5. Grahnén, A., C. Bahr, et al. (1979). "Bioavailability and pharmacokinetics of cimetidine." European Journal of Clinical Pharmacology **16**(5): 335-340.

6. Hoffman A, Z. E. (1997). "Pharmacokinetic considerations of new insulin formulations and routes of administration." Clin Pharmacokinet. **33**(4): 285-301.

7. Huang, W., S. Lee, et al. (2009). "Mechanistic Approaches to Predicting Oral Drug Absorption." The AAPS Journal **11**(2): 217-224.

8. Jantratid, E., S. Prakongpan, et al. (2006). "Feasibility of Biowaiver Extension to Biopharmaceutics Classification System Class III Drug Products: Cimetidine." Clinical Pharmacokinetics **45**(4): 385-399.

9. Marshall, S., F. Macintyre, et al. (2006). "Role of Mechanistically-Based Pharmacokinetic/Pharmacodynamic Models in Drug Development: A Case Study of a Therapeutic Protein." Clinical Pharmacokinetics **45**(2): 177-197.

10. Martinez, M. a. and Rilyn (2002). "A Mechanistic Approach to Understanding the Factors Affecting Drug Absorption: A Review of Fundamentals." Journal of clinical pharmacology **42**(6): 620-643.

11. Narasimhan, B. and N. A. Peppas (1997). "Molecular analysis of drug delivery systems controlled by dissolution of the polymer carrier." Journal of Pharmaceutical Sciences **86**(3): 297-304.

12. Ni, P. F., N. F. H. Ho, et al. (1980). "Theoretical model studies of intestinal drug absorption V. Non-steady-state fluid flow and absorption." International Journal of Pharmaceutics **5**(1): 33-47.

13. Salama, N. N., N. D. Eddington, et al. (2006). "Tight junction modulation and its relationship to drug delivery." Advanced Drug Delivery Reviews **58**(1): 15-28.

14. Somogyi, A. and R. Gugler (1983). "Clinical pharmacokinetics of cimetidine." Clinical

Pharmacokinetics **8**(6): 463-495.

15. van de Waterbeemd, H. and E. Gifford (2003). "ADMET in silico modelling: towards prediction paradise?" Nat Rev Drug Discov **2**(3): 192-204.

16. Willmann, S., A. Edginton, et al. (2007). "Development and Validation of a Physiology-based Model for the Prediction of Oral Absorption in Monkeys." Pharmaceutical Research **24**(7): 1275-1282.

17. Willmann, S., W. Schmitt, et al. (2003). "A Physiologic Model for Simulating Gastrointestinal Flow and Drug Absorption in Rats." Pharmaceutical Research **20**(11): 1766-1771.

18. Willmann, S., W. Schmitt, et al. (2004). "A Physiological Model for the Estimation of the Fraction Dose Absorbed in Humans." Journal of Medicinal Chemistry **47**(16): 4022-4031.

19. Yu, L. X. and G. L. Amidon (1999). "A compartmental absorption and transit model for estimating oral drug absorption." International Journal of Pharmaceutics **186**(2): 119-125.

20. Yu, L. X., E. Lipka, et al. (1996). "Transport approaches to the biopharmaceutical design of oral drug delivery systems: prediction of intestinal absorption." Advanced Drug Delivery Reviews **19**(3): 359-376.

(717c) Critical Quality Attributes (CQAs) of Biodegradable Polymer Matrices and Particles: Impact of a Recent Mathematical Model

Sam N. Rothstein, Department of Chemical Engineering, University of Pittsburgh, Qrono Inc., Pittsburgh, PA and Steven R. Little, Department of Chemical and Petroleum Engineering, University of Pittsburgh, Pittsburgh, PA; Bioengineering, University of Pittsburgh, Pittsburgh, PA; Immunology, University of Pittsburgh, Pittsburgh, PA; McGowan Institute for Regenerative Medicine, Pittsburgh, PA

Long-acting controlled release formulations address the top causes of non-compliance, a problem responsible for 10% of hospitalizations and over $100 billion in avoidable annual medical expenses. These formulation are also among the most difficult and costly formulations to design and produce because of their duration, which can extend to weeks or months, and the complexity of their drug-polymer compositions, which makes them difficult to model. Efforts to expedite the development and quality assurance of these formulations currently include temperature accelerated release testing methods and system specific transport or stochastic models. Methods or models with great efficiency, flexibility, and predictive power are desirable for the platform evaluation of polymer matrix-based controlled release formulations.

Researchers at the University of Pittsburgh have recently developed a robust model of polymer matrix controlled release and applied it to more than 50 formulations. Specifically, CQAs were evaluated in polyester and polyanhydride microparticles, nanoparticles, large matrices (pellets, rods, discs, tablets) and thin films loaded with agents ranging from small molecules of >300Da up to large protein complexes, and even viruses. These evaluations have identified five key input parameters: matrix geometry, spatial drug distribution, polymer molecular weight, polymer degradation rate and encapsulated drug size. Variations in these model inputs have been analyzed for their impact on drug release from microparticles loaded with the antipsychotics risperidone and quetiapine. Model predictions have also proven useful for the interpolation and extrapolation of in vitro release assays where data collection has been conducted infrequently or terminated early. Such predictions have confirmed equivalence of two protocols for the detection of enfuviritde release (anti-HIV) from poly(lactide-co-glycolide) microparticles. In another example, the simulations successfully extrapolated upon preliminary release data from a potentially unstable fatty-acid drug candidate and helped confirm the utility of an early microparticle formulation design. Software and formulation design services based on this model are currently being commercialized by Qrono Inc. www.Qrono.com

(717d) Effect of Acetate and Succinate Substitution Levels On Spray Dried Dispersions of Hypromellose Acetate Succinate – A Quality by Design Approach

William Porter III, Pharmaceutical Excipients, Dow Wolff Cellulosics, The Dow Chemical Company, Midland, MI, Nick Grasman, Dow Wolff Cellulosics, The Dow Chemical Company, Midland, MI, Oliver Petermann, Dow Wolff Cellulosics Gmbh&Ohg , Bomlitz, Germany, Steven Guillaudeu, Formulation Science , The Dow Chemical Company , Midland, MI and Meinolf Brackhagen, Dow Wolf Cellulosics GMBH & CO. OHG, Bomlitz, Germany

Pharmaceutical pipelines are broadly hindered by poorly soluble compounds and a need exists to facilitate solubilization to enable the use of a greater number of drug candidates. To meet that need, hypromellose acetate succinate is frequently used to produce amorphous solid dispersions, typically by spray drying the polymer with a drug candidate, resulting in supersaturated levels of drug compounds in solution. Currently, there are only three commercial grades of hypromellose acetate succinate available limiting the ability to tailor the drug release rate, supersaturation level and the extent to which recrystallization is prevented after drug release.

The present study utilizes a quality by design (QbD) hypromellose acetate succinate polymer set in which the acetate and succinate levels across the design space are varied. The samples in the QbD sample set are used to create spray dried dispersions with poorly soluble model drug compounds with varied physical properties. Evaluation of the performance parameters such as solubility enhancement, dissolution rate, and crystallization inhibition of the solid dispersions formed will be discussed. Furthermore, the advantages of utilizing a QbD approach to polymer substituent levels to tailor the performance of a drug candidate will be examined.

(717e) Hydrodynamic Effects of an Arch-Shaped Fiber Optic Probe in a USP Dissolution Testing Apparatus 2

Yiran Zhang, Otto H. York Department of Chemical, Biological and Pharmaceutical Engineering, New Jersey Institute of Technology, Newark, Gerard Bredael, Merck & Co., Inc. , Summit and **Piero Armenante**, Otto H. York Department of Chemical, Biological and Pharmaceutical Engineering, New Jersey Institute of Technology, Newark, NJ

Dissolution testing is widely used in the pharmaceutical industry to evaluate newly developed drug formulations and as a quality control method to insure that solid dosage forms have consistent dissolution property. Typically, samples are manually drawn from the dissolution vessel prior to analysis. An approach to overcome the limitations of manual sampling consists in the use of sampling probes, such as fiber optic probes, permanently inserted in the dissolution medium and continually sampling the drug concentration in it as the solid dosage form dissolves. Despite their advantages, permanently inserted fiber optic probes can alter the normal fluid flow within the vessel and produce different dissolution testing results. In this study, the hydrodynamic effects introduced by an arch-shaped fiber optic probe in a USP Dissolution Testing Apparatus 2 are studied by: (1) conducting dissolution tests, with and without the probe, using Prednisone tablets fixed at nine different locations at the bottom of the vessel and comparing the dissolution profiles obtained using statistical tools; and (2) experimentally determining the velocity profiles in the vessel, with and without the probe, using Particle Image Velocimetry (PIV) and quantifying changes in the flow velocities on selected horizontal iso-surfaces. The results show that the arch shaped fiber optic probe does have a baffling effect on the hydrodynamics in the dissolution vessel. This effect results in changes in the velocities in the fluid flow, and therefore in changes in the dissolution rate of the tablets undergoing testing. The baffle effect is observed mainly in the region where the probe is inserted. However, this perturbation is also found to reach the region below the impeller and to change the velocity profile there, resulting in differences in dissolution profiles when the tablets are fixed at positions that are downstream of the probe and within the low velocity region below the impeller. On the other hand, the hydrodynamic effect generated by the probe does not appear to be particularly strong. In most dissolution testing runs, the changes in dissolution profile are not large enough to fail the tests, according to the FDA criteria (f_1 and f_2 values). The PIV measurements additionally show that the baffle effect is not strong enough to break the overall flow pattern, or to affect the region around the impeller, which is dominated by the main flow generated by the impeller. It can be concluded that the hydrodynamic effects generated by the arch-shaped fiber optic probe are real and observable, resulting in slightly modification of the fluid flow in the dissolution vessel and therefore in detectable differences in the dissolution profiles. However, these effects are limited and do not typically lead to dissolution testing failures.

AUTHOR INDEX

AUTHOR INDEX